The Banana

At Table | University of Nebraska Press: Lincoln and London

The Banana

Empires, Trade Wars,
and Globalization

James Wiley

Photographs © 2008 by James Wiley. Maps ©
2008 by Matthew Craig. © 2008 by the Board
of Regents of the University of Nebraska. All
rights reserved. Manufactured in the
United States of America ♾
Library of Congress
Cataloging-in-Publication Data
Wiley, James.
The banana: empires, trade wars, and
globalization / James Wiley.
p. cm. — (At table)
Includes bibliographical references and index.
ISBN 978-0-8032-1577-1 (cloth: alk. paper)
ISBN 978-0-8032-3285-3 (paper: alk. paper)
1. Banana trade. 2. Banana trade—Latin
America. 3. Banana trade—Caribbean Area. I.
Title. II. Series: At table series.
HD9259.B2W52 2008
338.1'74772—dc22   2007042152
Set in Dante.
Designed by Ashley Muehlbauer.

*This book is dedicated to the
independent banana farmers of
the eastern Caribbean and to the
workers of Latin America's
banana plantations.*

Contents

Illustrations

Photographs

Maps

Tables

Preface

Late twentieth- and early twenty-first-century trends toward the continuing integration of the world economy are attracting the attention of geographers and other academicians who seek to assess the impacts that globalization processes have at various geographic scales. As new associations, such as the North American Free Trade Agreement (NAFTA) or the Common Market of the South (MERCOSUR), were created and others, like the European Union (EU), were broadened and deepened, the significance of trade in the global economy increased dramatically. This growth enhanced the impact of trade on places, people, and individual industries, affecting the path of development in various regions of the world. That such events occurred against the backdrop of a vastly changed institutional framework resulting from the conclusion of the Uruguay Round of the General Agreement on Tariffs and Trade (GATT) in 1994 and the formation of the World Trade Organization (WTO) in 1995 only served to exacerbate and geographically disperse their impacts. This led to the alteration of many long-established trade patterns, including some like the Lomé Convention that linked the developed states of the global North to the less-developed countries (LDCS) of the global South.

Agriculture is one of the sectors most affected by these changes. It is certainly true that agribusiness firms from developed countries were penetrating the production systems of the LDCS from the early years of the twentieth century and that their involvement in the South expanded steadily during the post–World War II era. Nevertheless, prior to the conclusion of the Uruguay Round of talks, most international agricultural trade flows were still characterized by high levels of protectionism or were subject to other regulatory frameworks that managed the direction and volume of trade in primary sector commodities. Thus, agriculture did not experience the slow but steady trade liberalization affecting much of the manufacturing sector following the implementation of the initial GATT system in 1947. That situation changed after 1994, with potentially dramatic effects on individual commodity industries. Since these developments are still quite recent, research in the form of individual industry case studies can prove useful in analyzing the impact of the developments on the various national and subnational regions where they are produced and marketed, and on the transportation industries that link those places. Such studies fall within the realm of economic geographers.

The banana is the world's most important fresh fruit commodity, at least when measured by volume of trade. The global banana industry is now a little more than a century old, having appeared on the scene in the late 1800s as a result of technological advances like refrigerated shipping, which facilitated the transportation of highly perishable goods to distant markets. Since its beginning, the banana industry has been fraught with controversy, exhibiting many of the issues that first emerged in economic relations between North and South during the era of European exploration of the non-European world.

Perhaps more than any other agricultural product, the banana reflects the colonial, neocolonial, economic nationalism, and contemporary neoliberal stages of the evolution of the world economy. At each stage, the greater changes occurring in the global economy manifested themselves in the economic geography of banana production and trade. This remains true as neoliberal imperatives drive the globalization process and mandate free(r) trade, influencing the patterns of transatlantic banana flows today. Specifically, the creation of the Single European Market (SEM) in 1992—itself a manifestation of the globalization trend—and the European Union's emergence as the world's leading market for bananas challenge the historic dominance of U.S.-based transnational corporations in the banana industry. This challenge led to the U.S.-EU banana war of the

1990s, placing the banana once again at the center of a major controversy. The war reached its peak when the United States responded to the European challenge by filing a complaint with the WTO, newly formed in 1995 with a mandate to remove unfair obstacles to the free movement of goods and services across international space.

This book is an attempt to demystify the banana trade dispute, presenting it as one event along the path toward the globalization of the banana industry. The book draws on the framework of the industry's historical economic geography in an analysis of the contemporary forces to which the industry must now respond. Its contribution to the rapidly growing body of literature on neoliberalism and globalization is to illustrate how individual economic sectors and the regions in which they function are differentially affected by globalization processes and, consequently, how they respond to it. Globalization is often presented as a monolithic and irresistible force, but the current situation in the banana industry indicates that this view is overly simplistic, disguising various efforts at resisting or attempting to modify the process at local and national levels. Nevertheless, the banana case does represent, in the final analysis, another triumph in globalization's march toward the greater integration of the world economy and increased penetration of neoliberalism in the agricultural sector. This conclusion is based on the responses of the EU to the banana dispute, the role played by the new WTO, the weakening of the Lomé Convention system, and the various efforts by small countries to overcome their relative competitive disadvantages in an era that emphasizes competitiveness in the international trading milieu. All of these subjects are discussed in the chapters that follow.

Barbara Welch (1996, 22) identified four subsystems within the global banana trade. These subsystems, each operating independently of the others, include the North Atlantic, the Western Pacific, Southern Africa, and Southeastern South America. This book focuses solely on the major portion of the first subsystem, the largest and most contentious of the four. This subsystem links Latin America and the Caribbean as two major banana-producing regions to the United States and Western Europe, the world's two most lucrative banana markets. The text does not consider the African portion of the North Atlantic subsystem, nor does it analyze the much smaller southeastern South America subsystem, which links Brazilian banana production to markets in Argentina and Uruguay.

The primary goals of this book are to provide an informative overview of the banana production and trade system involving Latin America and the Caribbean, and to serve as a case study that illustrates how any analysis of the impacts of globalization on a given industry must address the industry's past. That past must be analyzed to allow an understanding of how it shapes the industry's response to contemporary neoliberal forces. If both goals are successfully accomplished, the reader will gain knowledge of one of the Western Hemisphere's most significant agricultural sectors and develop a basis for analysis that can be extrapolated to other industries as well.

Acknowledgments

Academicians doing research abroad quickly realize how dependent they are upon the knowledge, insights, and goodwill of individuals who reside in those places. This project could not have progressed without the help of countless people who shared their ideas, resources, and humor with me during my many months in the field. Everyone whose name appears in the list of personal communications in the references section of the book and several who are not listed there are meritorious of my thanks. Their willingness to meet with me offered much to my work that I could not have gained by reading reports, documents, articles, or books. They range from banana farmers and workers to WTO ambassadors and EU officials, and I am deeply indebted to all of them and wish to acknowledge their contributions to this effort.

In several countries, extraordinarily generous individuals willingly gave much of their time to assist me with this project. Long, often multiple conversations with them helped me conceptualize and reconceptualize my approach to the topic, as events unfolded and banana geography continued to change. First, in Costa Rica, where I initially ventured into the world of bananas, Martin Zúñiga of CORBANA was instrumental in providing much of my early "banana education," helping me to understand the basic elements of the industry. During my first visit to one of the ACP states involved in the dispute, Carey Harris, Director of Dominica's Diversification Implementation Unit, impressed upon me the challenges of diversification in countries with specialized economies. Zelie Appleton opened many doors during my visit to the EU in Brussels and was the person who demystified the EU banana import licensing program for me. In Grenada, Oliver Benoit of the Ministry of Agriculture offered a unique perspective on the problems faced by small econo-

xvi | *Acknowledgments*

mies, adding a new dimension to my work that was further developed through my pivotal interview with Dr. Kathy-Ann Brown of CARICOM's Regional Negotiating Machinery in Europe. I wish to convey my most sincere appreciation to them all.

The logistics of foreign research are always challenging, but several people helped in a variety of ways. Diana Kieswetter-Alemán offered access to the UPEB files in Panama while David Dunkley and Luigi Stendardo provided valued assistance during the many hours I spent in the WTO library and document center. Closer to home, Chris Matthews gathered many documents, articles, resolutions, and other materials for my use during several visits to the EU Delegation Office in New York. I also wish to extend a very special thank you to Mrs. Horsford, administrative assistant to the Permanent Secretary of Dominica's Ministry of Agriculture, who illustrated one of the few real advantages of small size in a globalizing world. Whenever I asked how I might contact someone for an interview, she simply picked up the phone, talked to the person in question using first names only, and set up the appointment for me! Words alone do not convey sufficient thanks for her efforts on my behalf.

Finally, I would like to express my appreciation to Matthew Craig, a geography student at Hofstra University, for preparing the maps that appear in this volume, and to my colleague Jean-Paul Rodrigue, for his valuable assistance at several junctures and for his patience with my lack of technological skills. Last, though not least, I wish to acknowledge the generous financial support of the Hofstra College Faculty Research and Development Fund. Without their ongoing assistance, I would not have been able to undertake several dimensions of this research.

Introduction

The banana was introduced to consumers in the United States in the late 1800s. This nutritious addition to the American diet became possible only after the invention of the refrigerated ship. Now taken for granted, refrigerated shipping permitted timely transport of highly perishable fruit from tropical regions in Latin America and the Caribbean to markets in North America. The popularity of the fruit grew, and for most of the twentieth century the United States was the world's leading market for bananas.

In late 1998, when the mainstream press in the United States began to carry stories about a brewing banana trade dispute between the United States and the European Union (EU), most people reacted with surprise. Headlines foretold likely U.S. sanctions against a variety of European imports, none remotely related to fruit of any kind. Americans were accustomed to hearing about trade disagreements over automobiles, steel, electronic equipment, and other big-ticket items. But over a simple banana? What could all the fuss be about? The United States is not an exporter of bananas, so why would it allow itself to be portrayed as an aggressor bent on the economic destruction of small, friendly Caribbean countries, as indicated in several of the published reports?

The answers to these questions are far from simple. The roots of the trade dispute that erupted in full force in 1999 lie in both the distant and the recent past, extending to the very origins of the modern banana industry that emerged at the turn of the twentieth century. The banana is the world's top-ranked fruit commodity when measured by volume traded, and the industry has great economic importance in many parts of the world. Millions of people, particularly in the global South, directly or indirectly depend on it for their livelihoods. Furthermore, banana production has a great impact on industries from which it draws inputs (e.g., cartons, fertilizers, and chemicals) and uses services (e.g., railroads, trucking, ships, and ripening facilities). Most of these are based in the global North. All of this renders the banana worth fighting over; the historical geography of its production and trade demonstrates that many powerful economic entities have vested interests in the industry that they will go to great lengths to protect.

Issues Underlying the Globalization of Banana Production and Trade

The banana trade dispute that prompted the research for this book lasted from 1993 until 2001, but its outcomes will have greater longevity. The dispute was precipitated by the announcement of a new banana importation policy that limited the quantities of Latin American bananas that could enter the EU, as described in chapter 6. The United States' interest in this development is explained by its status as the host country to three major transnational corporations (TNCs) that dominate the Latin American banana industry. As in many agricultural sectors, greater profits are generated during the trade stage than by production itself, and any potential loss of market access could prove damaging to the companies involved. The dispute led the United States and several Latin American exporting states to file a complaint against the EU policy with the World Trade Organization (WTO), which drew greater global attention to the matter.

Despite the problems inherent in researching a controversial subject while it is still evolving, the research seemed worth undertaking because it was quite clear that the dispute would show the effects of globalization processes on one of the hemisphere's major food-producing industries. Several important issues underlying the U.S.-EU trade dispute transcend the specifics of the disagreement, which are discussed in chapter 9, and serve as illustrations of the broader process of globalization, the neoliberal policy umbrella that guides it, and the penetration of the oft-labeled "globalization/neo-liberal project" into the agricultural sectors of many less-developed countries (LDCS).

The drive toward free trade is one of the primary issues underlying the banana trade dispute. The last two decades of the twentieth century were characterized by a significant reduction in the impediments to the movement of goods and services across international boundaries. The creation of trading blocks such as the North American Free Trade Agreement (NAFTA) and the Common Market of the South (MERCOSUR), together with the deepening of the EU, necessitated the development of common import/export policies by member countries that previously operated as individual markets. These policies extended into agriculture to an unprecedented degree, threatening the continuity of many long-term trading relationships, often neocolonial in nature, between developed countries and LDCS that focused on primary sector commodities. For some EU member states, those relationships included trade preferences offering guaranteed market access for goods exported by LDCS that did not compete with products of the EU itself.

The reduction or removal of trade barriers led to an increased emphasis on achieving competitiveness in world markets, in keeping with the theory of comparative advantage that is the bedrock of the capitalist trading system. The search for new means of achieving competitiveness can stimulate technological innovation leading to better, cheaper products. In many cases, however, the technological innovations are too expensive for local producers in the LDCS. Nevertheless, the producers must purchase the technologies to avoid falling behind in the competitiveness race.

Competitiveness is often measured in quantitative ways that render the cost of production the most important variable. Quality factors may be undervalued by such measurements. In situations where labor is one of the major inputs into the production process, achieving competitiveness can be accomplished through the often-noted "race to the bottom," in which maintaining cheap labor takes priority over enhancement of social welfare. In such cases, questions of fairness arise, leading many to suggest that fair trade, rather than free trade, should be the priority of the global trading system.

The current emphasis on comparative advantage and competitiveness increases the tendency to reward larger-scale producers that are able to generate higher economies of scale. Economies of scale are achieved by spreading the fixed costs of production over a higher volume of units produced, thus lowering the per-unit cost of the product. The inability of small-scale producers to achieve this level of production contributed to the decline of family farms in the United States and their ultimate displacement by large-scale agribusiness. Similar threats now exist in the export agriculture sectors of many LDCs, into which the large food transnationals have moved with increasing frequency in the last two decades.

The issue of scale and the ability to generate economies of scale can be extrapolated beyond the individual producer to the national level. Small countries increasingly find themselves at a disadvantage within the new trading milieu. It appears unlikely that countries like Dominica or Cape Verde could ever expect to compete directly in the realm of production or transportation with countries like Brazil or South Africa. One may legitimately question what role can be played by the world's mini-states in an emerging world economy that is being shaped by neoliberal trade policies.

Finally, the role of vertical integration must be considered. Vertical integration is a model of industrial organization in which one firm controls several, if not all, stages of the industry. Thus, an agricultural company may manufacture various inputs used in crop cultivation or harvesting, operate farms or plantations, own the transportation systems used to bring the produce to market or to a processing plant, operate the factory, and so on. This is particularly relevant to the banana industry, which served as an early example of the vertical integration model on an international scale. Today, modified systems of vertical integration continue to characterize banana production and marketing, ensuring that control of the industry remains in the North.

The geographic realm of the banana industry discussed in this book extends to an array of producing and consuming states. The Latin American countries involved include five of the world's major banana exporters—Ecuador (the largest), Costa Rica (usually second), Colombia (third), Panama, and Honduras—along with Guatemala. These are the primary "third-country" sources of bananas consumed by EU citizens, meaning that they are not internal EU producers, nor are they linked to the EU through the Lomé Conventions. Their involvement explains the United States' interest in the banana trade because, collectively, they comprise the "dollar-zone" banana producers under the domination of U.S.-based transnationals.[1] Nicaragua, Venezuela, and Mexico, as lesser exporters whose economies focus on other primary sector commodities, such as coffee and oil, played smaller roles in the dispute and will not be discussed.

Eight Caribbean states are among the African, Caribbean, and Pacific (ACP) banana exporters to EU members. The ACP states are former colonies of EU members that are signatories to the Lomé Conventions, the first of which was signed in 1974. Seven are directly engaged in the dispute, which, for several, is a true economic crisis. The seven include five island nations: Jamaica and the four Windward Island countries of Dominica, Grenada, St. Lucia, and St. Vincent and the Grenadines (hereafter, St. Vincent). The other two are mainland states: Belize on Central America's Caribbean coast and Suriname on South America's north coast. The eighth country, the Dominican Republic, is a more recent addition to the ACP group and was not included among the traditional banana exporters covered by the Lomé Convention. Six African ACP states are involved: Cameroon, Cape Verde, Ivory Coast, Ghana, Madagascar, and Somalia. Somalia, whose primary notoriety in North America stems from its famine and political crisis during the early 1990s, exported bananas to Italy throughout that period. The African countries are beyond the purview of this book.

The European Union included twelve members when its controversial trade policy was implemented in 1993. Those were Belgium,

Denmark, France, Germany, Greece, Ireland, Italy, Luxembourg, the Netherlands, Portugal, Spain, and the United Kingdom. In 1995 Austria, Finland, and Sweden joined the organization and, by default, the banana fray. Four EU members are themselves banana producers, primarily in their overseas territories. These include the Canary Islands (Spain), Crete (Greece), Guadeloupe and Martinique (France), and Madeira (Portugal).

Collectively, these countries represent both the North and the South within the world economy, signifying differing levels of socio-economic development. They also vary significantly with regard to their capability to withstand the economic shocks that accompany the globalization process.

The Research Process

My initial exposure to the banana trade dispute occurred while I was in Ecuador as a participant in the Fulbright Program's South America Today seminar during the summer of 1993. With fellow grantees, I attended a presentation by Herman Van Sant, Filanbanco's liaison with the EU, at the bank's offices in Guayaquil on June 30, 1993. His subject was the new banana importation policy that, coincidentally, was to be implemented by the European Union the following day. As an economic geographer who specializes in both Latin America and the Caribbean, I was struck by the potential impact of this policy and its spatial ramifications in the Americas. Immediately, I began to plan a complete reorientation of my personal research agenda to accommodate this important issue. I never imagined that, more than ten years later, I would still be working on the subject, that it would be more complex and controversial than ever, and that its long-term resolution would remain inconclusive.

The research is based upon several kinds of source materials. Since 1993 I have conducted fieldwork on twelve occasions, not counting visits to the EU mission in Manhattan near my base at New York's Hofstra University. On each trip I interviewed people working directly in the industry and officials in its related institutions. I consulted many primary sources, including legal and other documents, published and unpublished reports, and statistical compilations. Finally, personal observations of the various banana landscapes that I was privileged to visit contribute to the descriptive aspects of this volume.

The journey has taken several turns and offered numerous surprises since that meeting in Guayaquil. My initial expectations and concerns that the new EU policy would spell disaster for Latin American exporters were reinforced during a 1994 visit to Costa Rica. There, all predictions indicated doom and gloom, although at a point in time when the impacts of the policy were too new to be effectively evaluated. Although subsequent accords would later yield statistics that refuted those predictions, I nevertheless experienced a very steep learning curve while in Costa Rica. As that was my first venture into the world of bananas, I needed to acquire a lot of information about the industry in its Latin American settings before beginning any attempt at analysis. It was quite clear, however, that whatever effects the new policy might have in Costa Rica or elsewhere in Latin America, they would be concentrated in those countries' distinct "banana zones." Therefore, I focused on the characteristics of these highly specialized geographic regions in my preliminary evaluation of likely outcomes. A second important subject of investigation was the diplomatic aspect of the growing dispute and the initiatives taken by the Costa Rican government at that early stage of events. I returned to Latin America in 2000, visiting Panama, the host country for the Union of Banana Exporting Countries (UPEB), an intergovernmental organization (IGO) of Latin American banana-producing states. The UPEB document center was useful as a source of data and background information about the industry and the escalating trade dispute.

In between those Latin American visits, I shifted my attention to Caribbean exporting countries, where I expected to find strong support for the EU policy. To learn about the insular Caribbean perspective on the policy, I visited Dominica in 1995 and 1996, Grenada in 1997, and St. Lucia in 2000. I initially focused on the very different nature of banana production in those countries, as compared to the Latin American system, and how the fortunes of the banana industry affected entire national societies in the eastern Caribbean. I discovered support for the EU policy, as anticipated, but I also found concern bordering on paranoia about the U.S. reaction to it. The Caribbean response revealed broader concerns about globalization processes and the future of small states. Their fear was instigating frantic efforts to achieve economic diversification in the countries I visited. A desire to learn more about this stimulated a slight detour in my investigations so that I could analyze the role of those diversification programs as part of the broader response to the globalization of agriculture.

I also visited Belize (1998) and Suriname (2001), two mainland countries incorporated into the Caribbean group within the framework of the banana trade negotiations. In Belize I focused on how the unique past of the country's banana industry shapes its perspective on the trade dispute. Belize's banana industry differs significantly from both the Latin American and the Caribbean models, and its recent transition from being a state-run sector to a privately held industry merited specific attention for this process. Suriname was the only country involved in the dispute that maintained a state-controlled banana industry. As a result, it had a different outlook on the dispute; its banana sector is motivated by goals that go beyond the mere generation of profits. This places Suriname in a difficult position with regard to the competitiveness imperative currently driving the industry, providing an impetus for it to carry on its fight to preserve market access in the EU.

Finally, amidst the several visits to banana producing countries, I visited two European sites where much of the activity related to the trade dispute occurred. In 1996 I went to Brussels, Belgium, the headquarters of the European Commission, the executive branch of the EU. There, at the source of the controversial policy, I focused on Directorate General VI, responsible for agricultural affairs. The EU perspective, as expected, differed substantially from what I had learned in Latin America and the Caribbean, particularly with regard to the new licensing system for importing bananas into the EU. It also differed substantially from my own initial assessment of the policy. In 1993–94 I predicted that the new policy would instigate a shift in the geography of banana production in the Western Hemisphere—a shift toward ACP exporters and away from Latin America. The EU argued that such a shift would not occur and that its policy was designed to prevent the reverse from happening—a shift toward Latin America and away from the Caribbean and Africa. Their rationale, therefore, was that their policy preserved the geography of banana production. Furthermore, the EU maintained that the U.S. challenge to its policy would lead to the demise of banana production in many Caribbean countries and generate tremendous displacement among the people affected. That U.S. challenge led me in 2001 and 2003 to the WTO headquarters in Geneva, Switzerland, where I interviewed representatives directly engaged in the negotiations.

Abbreviations

ACP	African, Caribbean, Pacific (signatory powers to the Lomé Convention)
APEB	Association of European Banana Producers
BFA	Banana Framework Agreement
BGAS	banana growers associations
CAP	Common Agricultural Program (of the EU)
CARICOM	Caribbean Community (originally the Caribbean Community and Common Market)
CBEA	Caribbean Banana Exporters Association
CEPAL	Economic Commission for Latin America and the Caribbean
CIA	Central Intelligence Agency
CSO	Central Statistics Office
DOM	departementes d'outre mer (overseas departments of France)
DSB	Dispute Settlement Board (of the WTO)
DSU	Dispute Settlement Understanding (WTO)
EC	European Community
ECU	European Currency Unit
EDF	European Development Fund
EEC	European Economic Community
EPA	Economic Partnership Agreement
EU	European Union
FAO	Food and Agriculture Organization (UN)
FCFS	first come, first served
GATS	General Agreement on Trade in Services
GATT	General Agreement on Tariffs and Trade
IGO	Intergovernmental Organization
IICA	Inter-American Institute for Cooperation on Agriculture

IMD	International Institute for Management and Development
IMF	International Monetary Fund
IRCA	International Railways of Central America
ISI	import substitution industrialization
IUCN	International Union for the Conservation of Nature
LDCS	less-developed countries
MFN	most favored nation
NAFTA	North American Free Trade Agreement (United States, Canada, Mexico)
NERA	National Economic Research Institute
NGOS	nongovernmental organizations
NTAE	nontraditional agricultural export
OAS	Organization of American States
OECD	Organization for Economic Cooperation and Development
OPEC	Organization of Petroleum Exporting Countries
SEM	Single European Market
SLBGA	St. Lucia Banana Growers Association
STABEX	Stability of Exchange (EU-ACP mechanism)
SYSMIN	System for Minerals (EU-ACP mechanism)
TNC	transnational corporation
TPRM	Trade Policy Review Mechanism
TRIPS	Trade-Related Aspects of Intellectual Property Rights
TRQ	tariff-rate quota
UB	United Brands (successor to UFCO)
UFCO	United Fruit Company (the United)
UNCTAD	United Nations Conference on Trade and Development
UPEB	Union of Banana Exporting Countries
US/LEAP	United States/Labor Education in the America's Project
USTR	United States Trade Representative
WINBAN	Windward Islands Banana Growers Association
WIBDECO	Windward Islands Banana Development Corporation
WTO	World Trade Organization

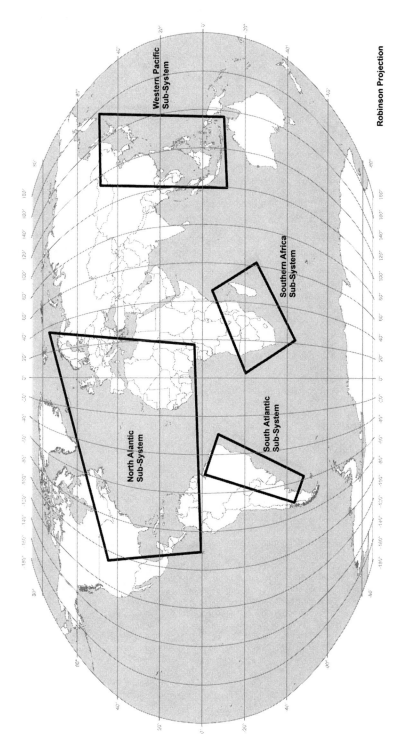

1. The four subsystems of the global banana trade

ONE

A Banana Plantation Model
Emerges in Latin America

1

The Creation of the
Banana Empire,
1900–1930

Contrary to popular belief, bananas are not native to the Western Hemisphere. According to Reynolds (1927, 19), their original homeland was most likely in South or Southeast Asia. Various banana species were diffused by human migration—for example, by the Polynesians, who carried them as far as Hawaii—or by conquest—such as by the Arabs, who were responsible for the banana's westward expansion. The biological term used for bananas, *musa*, is Arabic; it is a derivative of a Sanskrit word, reflecting the fact that Arabs originally encountered the fruit in India (Reynolds 1927, 23). There are many species of bananas, most of which are produced on a small scale for local consumption throughout the tropics. Only a few species are produced for export on a major scale, and those have lost the ability to reproduce from seeds. They must be propagated by dividing the rhizomes (root stock) to generate new plants for each growing season.

Europeans first came across bananas during their explorations of coastal Africa, to which the banana had spread either by Arab contact from the North, through Polynesian migration to Madagascar, or both. The Spanish were responsible for its introduction into the Americas when Friar Tomás de Berlanga brought the first banana plants from the Canary Islands to Santo Domingo in 1516. It is important to note that the fruit alone was incapable of such lengthy voyages, a problem that delayed its commercialization until the late nineteenth century. The entire plant stock had to be transported for propagation in the new territories.

This chapter focuses on the first stage in the development of the banana industry in the Western Hemisphere, drawing heavily from several early studies of the industry. Technological development was central to the development process, and the implementation of key innovations was fostered by the peripheral nature of the banana-growing regions within the countries in which the industry emerged. Distinct patterns of trade also were created, promoted by the United Fruit Company (hereafter, the United or UFCO), and the North and South became linked in a manner that has proven remarkably durable.

The Banana "Problematic"

Since its inception, the banana industry in the Western Hemisphere has been an early example of globalization. The industry was shaped by the requirements of the fruit itself, some of which do not differ very much from those of other fresh fruit or vegetable commodities. Other aspects of the industry, were unique, however, presenting special problems to those who sought to grow and market the fruit. The banana is highly perishable, with just a short period of time between its edible maturation point and the time of spoilage. Thus, timing is of the utmost importance to the banana industry. It is essential to harvest the still ripening fruit, transport it to a port facility, load it onto a ship, carry it to markets, complete the ripening process, and transfer it to wholesalers and local retailers in a coordinated manner. The span of time from harvest to consumption is only four to five weeks, during which the banana may have to travel up to fifteen hundred miles by land and from two thousand to six thousand miles by water to reach its market, where its retail shelf life will be no more than five days. While most major market regions lie in the temperate zones of the world, cultivation of the banana is limited to tropical or subtropical areas that have warm year-round temperatures and receive at least sixty inches of rainfall annually (OAS 1975, 5).

Therefore, the geography of banana production and consumption is determined by the fruit's need for high levels of efficient organization. This can take a variety of forms. But one thing has always been clear with bananas: it is virtually impossible for independent farmers, operating alone, to exercise sufficient control to ensure the eventual timely delivery of the fruit to the market. At a minimum, some form of association is needed, if not one or more corporate structures. This reality contributed to the emergence of the so-called Banana Empire. Ellis (1983, 35) identifies the banana industry's first stage as the period from 1870 until 1898, during which independent farmers in Central America sold fruit to private North American shipping firms. While this activity established that there was a market for bananas in the North, the period represents the banana's pre-industrial stage. The commercial banana industry really began with the founding of the United Fruit Company (UFCO) in 1899.

The process that ultimately brings bananas into consumers' kitchens requires numerous inputs, many specially designed facilities, and an institutional framework. At a minimum, the inputs needed include plant stock, land, labor, capital, fertilizers, pest control mechanisms, and cartons. The latter three are most affordable when purchased in bulk, which encourages the formation of associations in areas where small-scale farming is the norm. Investment capital merits special mention because of its controlling influence over all of the other inputs. Infrastructure requirements include packing facilities on or near the cultivation site, roads or railroads to link farms to seaports, ships with refrigerated chambers during international voyages, and ripening facilities in market countries. Institutional needs include agricultural credit, disaster relief, marketing, and research systems, all of which are instrumental in helping farmers sell their produce, ensuring the survival of the industry. The research needed to improve pest control technologies and develop disease-resistant fruit varieties is also beyond the capacity of individual farmers.

Thus, the production and marketing of bananas is a complex affair. Ellis (1983, 19) notes that it differs from most other agricultural sectors in the degree to which it is both heavily capitalized and reliant upon large infusions of labor. It requires a coordinated organizational framework that links cultivation areas in tropical regions with consumers mostly living in the temperate climate zones. The capital needed to accomplish this during the industry's incipient stage was only available in the North; therefore, control over the industry historically resided in the consuming regions rather than in the producing regions. This uneven relationship has been problematic from the beginning.

The Banana–Railroad Connection

The early commercialization of the banana and the origins of the banana trade in the Western Hemisphere are closely linked to the development of transportation infrastructure in Central America and Colombia. The territories of Costa Rica, Colombia, Guatemala, Honduras, and Panama were poorly internally integrated at the point in the late nineteenth century when refrigerated shipping made the banana trade possible. This was particularly apparent in the Caribbean regions of those countries, which lay far from the historic core regions in the highlands. The poor integration was relatively unimportant until economic liberalism (i.e., more open economies with fewer barriers to trade) gained ascendancy in much of Latin America during the latter half of the century. After 1880, governments became interested in creating more effective transportation networks to link inland population centers to the Pacific and

Caribbean coasts, and foreign experts and sources of capital were invited in to build the desired railroads. The foreigners invited into Central America were more likely to be from the United States than from the United Kingdom (UK), although the British had built the railroads of South America. The lack of national integration presented an excellent opportunity for U.S. capital to penetrate the region as never before.

A key event in the history of the hemispheric banana industry was the 1884 signing of the Soto-Keith contract between Costa Rican secretary of state Bernardo Soto and U.S. businessman Minor Keith. The contract obligated Keith to construct the remaining fifty-two miles of a railroad begun during the 1870s between the Caribbean port of Puerto Limón and San José, the capital of Costa Rica, which is located four thousand feet above sea level in the interior of the country. The earlier project had been financed by British capital under unfavorable terms that led to the accumulation of a substantial national debt. Keith renegotiated that debt, which Costa Rica would guarantee with export revenues. The country's primary motivation to complete the railroad was to more efficiently ship coffee, not bananas, to its European markets. Previously, coffee had been exported through the Pacific port of Puntarenas, necessitating a long voyage around Cape Horn. A railroad to the country's Caribbean coast would reduce that journey by more than half.

In return, the contract provided Keith with grants of eight hundred thousand acres of undeveloped national land, amounting to 7 percent of the national territory. He was obligated to improve the land, most of which was located in the country's sparsely settled Caribbean lowlands. The deal included a twenty-year land tax exemption, guaranteed the use of the country's railways for ninety-nine years, and permitted the duty-free importation of materials needed to build the railroad (Government of Costa Rica 1884/1989, sec. 22). This contract became the prototype for subsequent concessions and established the tone for future agreements that allowed the penetration of foreign capital into Caribbean Central America.

The Costa Rican railroad project was difficult and expensive, taking many years to complete. The route scaled a rise of more than four thousand feet from the Caribbean plains to the Meseta Central, Costa Rica's colonial hearth. Keith encountered many difficulties, which were later confronted during the construction of the Panama Canal. Nearly four thousand people, including three of Keith's brothers, died during the project, mostly from malaria and other tropical ailments. The majority of the casualties were English-speaking Afro-Jamaicans brought in for the project (Reynolds 1927, 41–42).[1] Keith realized that he needed to generate capital to finance the construction, so he began to develop banana plantations on the land granted to him in the contract, essentially starting Costa Rica's banana industry as an adjunct to the railroad project. Once construction was completed, however, the situation reversed and the railroad came to serve the rapidly growing banana business (Kepner and Soothill 1935, 49). Keith negotiated additional contracts with the Costa Rican government in 1892 and 1894 to develop additional railroads, including a network of spur lines in the Caribbean region to link the expanding banana zone with the main line and Puerto Limón (Ellis 1983, 60). These were organized as the Northern Railway Company in 1900, with the resulting railroad network having far more effective penetration in the banana region than the national railroad provided for the rest of the country.

A similar pattern emerged in Guatemala, where coffee was also the major export during the late 1800s and there was no rail link to the country's Caribbean coast. The Liberal Party's dictatorial regimes that ruled Guatemala after 1873 embarked on a modernization program that simultaneously served the coffee-growing elite and used the export earnings from coffee to diversify the national economy. Their goals would be realized more quickly if they could use the more expeditious Atlantic route to ship coffee to European markets, freeing the country from near-monopoly domination by U.S. shipping firms that served Guatemala's Pacific ports. But the country lacked a satisfactory link between Guatemala City in the coffee-growing highlands and Puerto Barrios on the Caribbean coast. To overcome this gap, construction of the Northern Railway began during the 1880s, using a combination of British, German, and local capital, with much of the latter generated through the seizure of indigenous communal lands and the confiscation of church properties (Dosal 1993, 17). As in Costa Rica, engineering problems and the accumulation of debt during the project led to failure in 1885. Subsequent efforts to raise additional capital during the 1890s collapsed along with global coffee prices, causing the government, which previously resisted the penetration of U.S. capital, to turn to Minor Keith in 1900. A concession was signed that gave Keith's Central American Improvement Company the right to operate the Northern Railway for ten years, and to retain all profits from doing so, an exemption from export taxes, and land grants of 57,000 acres in the Caribbean lowlands. The land would be planted with bananas to generate freight for the railroad. In return, the company agreed to complete construction of the line (Dosal 1993, 40–41). This agreement was replaced by an even more favorable contract in 1904 in which Keith committed to building the last sixty-mile stretch of the Northern Railway and was authorized to build and acquire railroads throughout the country. The contract also provided him with 168,000 additional acres of land, a thirty-five-year exemption from export and local taxes, control over the docks at Puerto Barrios, and the right to operate virtually free of government interference in the company. Dosal (1993, 44) describes this concession as "incredible" and suggests that it could only have occurred through the corruption of key government officials. The company's name was changed to the International Railways of Central America (IRCA) in 1912, reflecting its expansion into other regions of Guatemala and nearby El Salvador, where bananas were not produced (Kepner and Soothill 1935, 155).

Elsewhere the pattern of railroad development more clearly reflected the enclave nature of the emerging banana lands and the early fruit companies' greater interest in building railroads to serve their own banana plantations than in developing nationwide transportation networks. In Honduras the government's attitude toward granting concessions in return for the construction of railroads passed through three stages during the first four decades of the twentieth century. During the first stage, its initial willingness to deal with the large fruit companies was motivated by its need for a national railroad network that linked Tegucigalpa, its inland capital, with the country's Caribbean coast and its desire to use the railroad system to stimulate agricultural development. Unlike other countries, Honduras was able to prevent a monopoly from developing. It negotiated with several fruit companies, allowing each to operate within a specifically defined region along the banana-producing northern coast. It granted land in return for railroad construction to the Vaccaro Brothers Company in 1906 and to the Cuyamel Fruit Company of New Orleans in 1912. In 1913 it granted land concessions of nearly four hundred thousand acres to two UFCO subsidiaries, the Tela Railroad Company and the Trujillo Railroad Company, in return for the construction of railway lines penetrating inland from the coastal ports bearing those names (Ellis 1983, 44). A second stage began as Honduran enthusiasm about land concessions waned following World War I, when the country realized that the great prosperity enjoyed by the companies was not shared by the nation at large (Kepner and Soothill 1935, 151). Honduras imposed a national tax of one cent per box on banana exports, with an additional half-cent to go to the municipality in which the bananas were produced (Ellis 1983, 67). A third stage was shaped by the realities of the Great Depression, when the transnational corporations (TNCs) demanded greater concessions while providing less in terms of railroad development. Ultimately, Honduran aspirations were thwarted by the failure of the fruit companies to construct the promised railroad lines. The contracts Honduras negotiated invariably included penalties for nonfulfillment. In the end, the companies found it cheaper to pay the fines than to build railroads in the highlands. Tegucigalpa, the capital, and other interior cities were never linked to the country's Caribbean coast, although that region was well served by railroads that joined several banana-growing districts with individual port cities.

In Colombia the UFCO assumed control of a British company that built the Santa Marta Railway, serving the port city of Santa Marta, the heart of the country's banana-producing region. The United subsequently built branch lines, linking them to the main line and thus facilitating shipment of bananas to the port. But the company did not fulfill the government's wishes to extend its railroads beyond the Santa Marta region to serve other parts of the country, and during a dispute in 1915 it successfully defended its concession against government efforts to assume control of the railroad (Kepner and Soothill 1935, 164).

In Panama the United was able to fill the power vacuum created by the political uncertainties at the turn of the century. Panama began the twentieth century as a rebellious province of Colombia, gained its independence in 1904, and immediately focused its attention on the construction of the Panama Canal. Meanwhile, the UFCO was able to operate without the formal contracts it needed elsewhere. In the Bocas del Toro banana-growing area along Panama's Caribbean frontier with Costa Rica, the company developed a network of railroads to haul bananas to the port at Almirante. Its use of the port was authorized through a 1904 law passed shortly after Panama's independence (Ellis 1983, 68). The region was not linked to the rest of the country, however, although Almirante was connected to the nearby Costa Rican Talamanca Valley and its banana plantations. Dosal (1993, 9) noted that it was easier to get to Almirante from Boston (early UFCO headquarters) than from Panama City. On the Pacific side, the company gained approval in 1926 for priority use of a state-constructed 34.5–mile railroad line connecting Puerto Armuelles to agricultural areas in the western Chiriqui province. The UFCO was granted control over use of the portion of the line extending from the banana zone around Progreso to the port. It could ship its fruit at inexpensive rates while bumping other cargoes and passenger trains that might interfere with the timely arrival of bananas in the port (Kepner and Soothill 1935, 167–68).

Although the details vary from one country to another, there is a remarkable similarity among all of these arrangements. In each case a foreign corporation was granted concessions to operate in a Latin American country while incurring relatively few obligations in the process. The small sizes of the countries involved, their even smaller relative populations, and the uneven distribution of those populations all facilitated the rendering of exploitative agreements by governments that lacked other options for developing their countries' remote regions.

The Founding of the UFCO and Horizontal Integration

The favorable railroad concessions and contracts in Costa Rica and Guatemala provided the basis for increased activity by U.S. entrepreneurs in the region. Access to land was a key element in this process. Nevertheless, despite the size of Minor Keith's landholdings in the two countries, there were limits to what he could accomplish on his own as a supplier of a commodity that was still relatively unknown to most North American consumers. Fulfilling the infrastructure needs of the banana industry was beyond the means of all but the largest firms. Therefore, in 1899 Keith joined forces with the Boston Fruit Company, a firm that had imported bananas into major northeastern U.S. ports from West Indian sources through seven subsidiaries since 1885.

The merger created the United Fruit Company, incorporated under the laws of New Jersey with authorized capital of twenty million dollars (Kepner 1936, 41). The result was a company whose size and geographic diversity ensured for the first time a permanent supply of bananas from the West Indies, Central America, and Colombia (Reynolds 1927, 51). Never before had one consolidated company been able to control such widely dispersed sources of the fruit. Production shortfalls due to storms in one region could be compensated for by additional supplies from other areas. Once established, the company expanded, clearing new areas of Caribbean rainforest in Central America. At the same time, it concentrated on increasing the demand for bananas in the United States and Europe.

A rapid increase in the volume of the banana trade followed the merger. Costa Rica's exports grew from 420,000 stems in 1884 to nearly 3 million in 1899 and more than 10 million by 1907. They peaked at more than 11.1 million stems in 1913 and then stabilized at an average of about 8 million per year following World War I (Kepner and Soothill 1935, 52). Similar patterns occurred in the other countries where the company did business. The four primary Central American exporters—Guatemala, Honduras, Costa Rica, and Panama—experienced a collective increase from 8,219,343 stems in 1900 to 43,332,224 in 1930. The Depression caused demand to decline, resulting in the export of just 24.7 million stems in 1935 (Ellis 1983, 51–53).

The company expanded quickly following the merger, growing in a manner that allowed it to achieve both horizontal and vertical integration. The nature of the UFCO's horizontal integration disguised its true scale because much of its growth occurred through subsidiaries that did not use the United Fruit name—a strategy that became a UFCO trademark. The subsidiaries were key players in the development of the empire. Several were automatically incorporated into the new system by the act of merger itself; Boston Fruit had operated seven subsidiaries in the West Indies, particularly in Jamaica. Keith himself was owner or part owner of three companies that became part of the empire, including the Snyder Banana Company, through which he acquired land in the Bocas del Toro district of Panama. One month after the merger, the UFCO purchased seven commercial firms that operated in Panama and Honduras (Ellis 1983, 43), and it later bought the Colombian Land Company, which held banana production land in the Santa Marta area, from its parent British firm (Kepner 1936, 39).[2]

In Costa Rica the Tropical Trading and Transport Company was organized to handle production and transport within the country's banana zone. Shortly after its founding, the UFCO began working to gain a monopoly over railroad transportation in Costa Rica, which was achieved when Keith's Northern Railway Company assumed the lease to operate the national railroad line (Kepner and Soothill 1935, 49). The railroads became one of many weapons in the company's arsenal used to dominate Costa Rica's banana industry.

Similar patterns evolved elsewhere, reinforcing the importance of railroads to the banana industry through the UFCO's dominance of most of the region's rail lines. The UFCO owned or controlled through advantageous lease arrangements the IRCA of Guatemala, the Tela and Trujillo Railroads in Honduras, Panama's Caribbean coast Changuinola Railroad, the branch lines serving UFCO plantations in all of Central America's banana producing regions, the Magdalena National Railway in Colombia, portions of the Chiriqui National Railway along Panama's Pacific coast, and the National Railroad of Honduras. Only a few lines, primarily in Honduras, were owned by rival firms. This gave the UFCO a near stranglehold on commercial transport systems. The company used this advantage to gain monopoly control over the banana industry in vast areas.

The subsidiary approach was also applied to the international transport stage of the industry. The UFCO purchased controlling interest in several shipping firms and placed its own representatives on the board of each company (Kepner 1936, 42). In 1904 it organized the Tropical Fruit Steamship Company Ltd. in the UK (Kepner and Soothill 1935, 180). In 1902 the UFCO associated with Elders and Fyffes Ltd., an Irish firm that marketed Jamaican bananas in the UK and later began to import Costa Rican bananas into Europe. By 1910 the UFCO had purchased all Elders and Fyffes stock shares, and the firm joined its growing family of subsidiaries, where it would remain until the late 1960s (Reynolds 1927, 56–57). Fyffes Ltd., as the company was later known, became one of the two largest British banana-importing firms.

An Empire Emerges: Land Acquisition

Land is an essential ingredient in all empires, and the Banana Empire was no exception. Following the merger that created the UFCO, land acquisition was one of the company's most important goals. In addition to the concessions gained through railroad construction contracts, the company used several other means to gain access to land. By 1930 UFCO-controlled tropical landholdings covered more than 3.4 million acres, an area greater than fifty-four hundred square miles, larger than Connecticut and any Caribbean island other than Cuba and Hispaniola. The historical economic and political geography of Central America's lowland tropics contributed to the relative ease with which the UFCO and its rivals acquired land. The politically fractious nature of the region, its historic poverty, the abundance of land relative to population size, and the uneven distribution of its population all contributed to this process. Cultural factors also had an impact. The mainstream of Central America's societies exhibited an aversion to living and working in the humid lowlands, which had seen little migration before the development of the banana industry. The region was, in essence, a frontier. All of this served to deflate land values in those areas and reduce the nationalistic resistance that might be expected with foreign control over land. As most of the land involved was unused, it was not viewed as a major loss.

Railroad concessions continued to serve as the primary means through which fruit companies gained control over land. In addition to the Costa Rican and Guatemalan cases already described, UFCO subsidiaries obtained more than 175,000 acres from such concessions in Honduras (Kepner 1936, 72). There, however, the land conceded to foreign companies was not granted in contiguous blocks. Instead, it was divided into large, alternating parcels of four thousand to five thousand hectares, with intervening parcels controlled by the government. This was done ostensibly to prevent total domination by a foreign company within a given region, but it also increased access to land by independent Honduran farmers (Kepner and

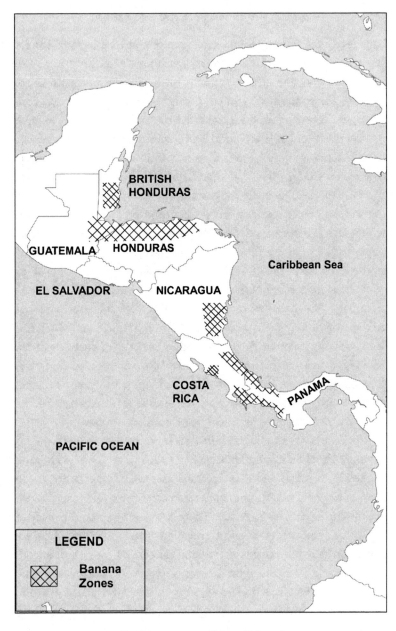

2. The banana zones of Central America

Soothill 1935, 114). The system worked against the development of the efficient plantation structure that the fruit companies sought to create. However, since parcels were available for leasing, the companies often found means to buy leases from the Hondurans who held them, thus effectively negating the intended spirit of the legislation.

In Costa Rica and Honduras, "denouncement" laws permitted people to claim unused national land. A person who wished to claim such a parcel of land had to file a denouncement with the local authorities. If no one disputed the denouncement, the land was made available for sale and would go to the highest bidder. In cases where the denouncer offered a bid equal to the highest received, he or she would be granted the right to purchase the property (Kepner 1936, 77–78). This process served the company well because there often was little competition in the bidding for denounced lands. Again, nationals were used as intermediaries to denounce land on behalf of the company. In other situations, citizens denounced land and purchased it under the assumption that they could resell it to the UFCO.

The use of the "gracias law" was another means through which the UFCO obtained substantial amounts of land in Costa Rica. This law was intended to provide a means of generating revenues for municipal governments. The gracias system allowed local authorities to dispose of their unused national land by selling the right of denouncement. A person buying the right could then denounce the land and subsequently bid for its purchase. The United, through its Gulf of Dulce Land Company subsidiary, used the gracias system to acquire nearly 350,000 acres of land along Costa Rica's Pacific coast before the law was terminated in 1926 (Kepner 1936, 79–80).

Finally, in many situations, access to land did not involve outright ownership. Renting or leasing of national lands not available for purchase could be arranged, often for lengthy periods of time, at very cheap rates. Such arrangements were common in Costa Rica, Honduras, and Guatemala.

Innovation: Technology and Vertical Integration

The time-sensitive nature of banana production and transport provided an advantage to a company characterized by vertical integration—that is, a company simultaneously controlling several stages of an industry. With vertical integration, a firm could avoid having to rely on the ability and willingness of other companies to deliver services, particularly transportation, in a timely manner in a situation where the failure to do so would spell disaster. In-house control over such matters was clearly beneficial. Much of the success of the United Fruit Company was derived from its vertically integrated organizational structure.

The UFCO rapidly developed its vertical integration structure, much of it through subsidiaries. Before 1900, independent producers raised most of the region's bananas. Once the UFCO acquired land, however, it engaged more directly in the cultivation of bananas and other crops such as cocoa. It began to purchase fewer bananas from local sources. This change increased the company's involvement in the internal political affairs of the countries in which it was operating. The operation of plantations also led to the company's diversification into ancillary activities, some of them income-producing activities such as the operation of commissaries, and others that could be viewed as employee benefits, such as housing and medical facilities. The need to pack bananas quickly at or near the production site dictated that packing facilities also be developed on the plantations.

The UFCO also gained access to, if not full control of, dock facilities at seaports such as Puerto Limón, Costa Rica; Puerto Barrios, Guatemala; Tela and Trujillo, Honduras; Puerto Armuelles and Almirante, Panama; and Santa Marta, Colombia. These ports ostensibly were open to the cargo of other companies, so controlling dock operations represented a substantial benefit to the UFCO in determining whose cargoes were to be loaded first. Over time the UFCO assembled its own fleet of ships to transport the fruit to ports in Europe and North America. This allowed the firm to diversify its operations by handling imports going into Central America and Colombia, transporting other goods exported by those countries, and even carrying the U.S. mail. Its ripening facilities in northern ports handled the fruit in its final stage of processing, before it was sold to wholesale and retail firms.

The importance of vertical integration is several-fold. Beyond the timing aspects already noted, a vertically integrated company is able to generate profits from each of the industry stages in which it is involved. In the case of bananas, this is critically important because, at best, only ten or eleven cents of each dollar spent by consumers on bananas go to the farm or plantation that cultivated the fruit. Often the amount is less, on occasion falling below five cents. The rest of the money goes to the various transportation companies that handle the fruit in both the producing and consuming countries, to ripening facilities, to wholesalers, and to retailers. The irony is that the grower bears the greatest risk of any of the operatives in the multistage industry.

Another benefit of vertical integration is that it enables a firm to affect what other participants in the industry can do. The UFCO was unparalleled among the fruit transnational corporations (TNCs) in the breadth and depth of its vertical integration, and it discovered that its control over several stages of the industry could prevent potential rivals from gaining viable footholds in several producing areas. Its willingness to resort to unethical practices, including the corruption of local and national officials to accomplish such goals, is the primary factor that, in time, made the company the foremost symbol of Yankee economic imperialism.

Control over railroads was a key to the process, as exemplified by the situations in Costa Rican and Guatemalan. When the UFCO-controlled Northern Railway assumed the lease for Costa Rica's national line, it gained the upper hand over potential rivals. It promulgated an order stipulating that any company wishing to transport bananas by rail had to provide a detailed requisition in advance. This alerted the UFCO to rivals' activities, enabling it to undermine them. The company had several means at its disposal for doing so. The Northern Railway altered train schedules to ensure that substantial quantities of fruit would spoil before reaching the port facility. It charged higher prices for the transport of fruit, causing financial hardship to rivals while merely necessitating bookkeeping transfers where UFCO fruit was involved. Such tactics made it difficult for other fruit companies to establish themselves in Costa Rica because they had no control over the timely shipping of their highly perishable fruit. The UFCO also entered into exclusive long-term contracts to purchase fruit from independent farmers, making it difficult for rival firms to do so (Kepner and Soothill 1935, 71–74).

In Guatemala, UFCO control of the IRCA allowed it to operate in a similar manner. Its rail freight tariffs bore no relationship to the distance covered. Coffee, Guatemala's leading export, provides a clear example of this. Nearly all of Guatemala's export coffee is grown in the highlands, just inland from the Pacific coast. Yet the company charged higher rates for coffee that it transported to Guatemala's Pacific ports, which it did not control, than it did to the far more distant Caribbean Puerto Barrios, which it did control (Kepner and Soothill 1935, 160). Coffee thus subsidized the rail network that also delivered bananas to Puerto Barrios. Since shipping is attracted to ports that offer the opportunity to handle cargoes in both directions, the ability of the UFCO to deliver the bulk of both major Guatemalan commodities to Puerto Barrios ensured that its port and railway would handle most of the country's import traffic as well.

The company's oceangoing vessels came to be known as the Great White Fleet. The ships, numbering approximately one hundred by 1930, essentially constituted the navy of the Banana Empire. They had refrigerated cargo holds to carry bananas as well as storage capacity for other commodities and goods. As an industrial carrier whose ships plied a variety of routes, the company could attract many other forms of cargo, including passengers. Due to the sensitive nature of their principal cargo, the UFCO's ships were guaranteed to sail on their predetermined schedules and move rapidly to avoid spoilage. Thus, the company could take advantage of many potential earnings opportunities.

The UFCO owned most of the ships it used, usually through its shipping subsidiaries, and chartered additional vessels as needed. Generally, though, it did not trust other fruit companies to ship its bananas—an indication, perhaps, of its own treatment of competitors' produce. More than half of its ships were used to transport fruit to North American ports, reflecting the United States' position as the leading consumer of bananas on the world market, but more than 40 percent of its ships served European destinations. Ironically, the company initially favored building its ships in the UK and flying them under the British flag, because costs were lower and regulations were more flexible there.

With its control of shipping, the United was able to develop creative pricing mechanisms that combined fares for its rail and boat services without divulging the component parts of the total fare. This allowed it to undermine competitors who operated railroads or ships but not both. Similarly, it raised its rates to punish those who might use its competitors' trains or ships. This occurred with the Pacific Railway in Costa Rica, which was operated by the government independently of the UFCO. In 1929 the Pacific decided to lower its tariffs for hauling coffee from the country's central highlands to the Pacific Ocean port of Puntarenas as a means of diverting freight from the UFCO-controlled railway serving Puerto Limón. The company responded by raising its steamship freight charges for coffee shipped out of Puntarenas by 30 percent, offsetting the benefit realized by the lower rail charges (Kepner and Soothill 1935, 185).

Operating commissaries proved to be a profitable ancillary activity in the UFCO's plantation areas. Managers of the commissary division were expected to produce a profit that could be used to offset labor costs in the company's other divisions. The company used several means to ensure that its commissaries were busy. It issued coupons as partial payment of wages to many workers; these could be exchanged only in UFCO commissaries. Employees also could obtain credit against future wages by using commissary order blanks. An installment buying system was instituted for more expensive items. To many, the commissaries represented a new form of debt peonage, a centuries-old Latin American tradition that bound rural workers to their patron in a continual cycle of debt. Indebtedness reduced the bargaining power of labor because it decreased the mobility of workers with outstanding bills. Strong-arm tactics, including deployment of local police, were employed to collect debts (Kepner and Soothill 1935, 322). Other merchant interests in the banana-producing regions were negatively affected by the company's commissary operations. They were unable to achieve the economies of scale that characterized the UFCO's stores, which derived the additional benefit of advantageous shipping charges for merchandise transported on company ships and trains.

The final elements of the United's vertical integration were found in the consuming countries. The United and its rivals operated ripening facilities in the port cities that received the shipments of bananas. The fruit generally required an additional three days for ripening before it was transported by rail or truck to wholesale and retail outlets. Because the distribution of bananas within market countries also was a source of profits, the competition among the fruit companies was fierce. The UFCO's major marketing subsidiary in the United States was the Fruit Dispatch Company. By the early 1930s the Fruit Dispatch Company had secured 80–90 percent of the U.S. market for the UFCO's bananas.

Clearly, the vertical integration strategy served the UFCO very well. Its rivals adopted similar frameworks, although at smaller scales and without the same degree of success. The United did not tolerate competition very well and continually sought to remove its rivals from the scene. Where other tactics failed, it simply tried to take over its rivals and incorporate them into the empire. In 1929 the Cuyamel Fruit Company, a major player in the Honduras banana trade, met this fate when the UFCO purchased controlling shares of its stock. This removed the last significant challenger to the United's dominance. As the 1920s drew to a close, the company enjoyed near-monopolistic control over the industry. With virtually all of the pieces of its empire in place, it could set about solidifying its position and focus on ruling its realm.

Splendid Isolation: The Banana Zone

The land base of the Banana Empire was not static. Rather, it shifted with changing conditions of production, such as soil exhaustion, storm damage, and plant diseases such as Panama Disease and Sigatoka negra. It also grew or shrank with changes in market demand for bananas that occurred with the onset of the Great Depression. Most of the empire's territory was found in the humid tropics of Central America and Colombia, occupying Caribbean lowlands. By 1930, however, problems with Panama Disease led to the acquisition of significant landholdings on Central America's Pacific side.

Remoteness was a unifying characteristic of most of the empire's territory on both the Caribbean and the Pacific sides. This trait came to be labeled "splendid isolation," signifying the beneficial aspects of the limited presence of state authority. This meant that the Banana Empire could operate with relative impunity within its territory. The United's functional autonomy enabled it to pursue its corporate policies without regard for how its practices might more widely affect whichever country was involved. The governments of the states to whom the empire's territory ostensibly belonged had not effectively penetrated the newly emerging banana zones. With the exception of Colombia, the countries' small size and low development level rendered their governments powerless in the face of the immense corporation that came to dominate sizable segments of their national territories. Indeed, agencies of those states were more likely to be pressed into the company's service than to challenge it directly. Pervasive poverty offered few development options for the areas under UFCO control, and low salaries for most public officials—still a

1. Central American plantation, a banana monoculture.
2. Plantation infrastructure in Costa Rica.

problem in Latin America—rendered them corruptible by company payoffs in return for concessions, reduced tax burdens, and other favors (Kepner and Soothill 1935, 209–21).

The region's banana zones, therefore, developed within this situation of splendid isolation. But what were these unique geographic entities really like? At its core, a banana zone is an excellent example of a monoculture, the cultivation of just one crop in a field, on land previously occupied by tropical forests of great biodiversity. Conversion from forest to plantation occurred with great environmental impact, a fact that has only recently gained appreciation as a result of the late-twentieth-century assault on rainforests worldwide. During the decades immediately after the creation of the UFCO, however, such changes in areas of sparse population were viewed as improvements. This was reflected in early writings about the industry (Reynolds 1927).

The land modification necessary to carve banana plantations out of a tropical forest required the marshalling of vast resources, providing yet another rationale for the involvement of a large corporation. In a region where export agriculture was dominated by medium-or small-scale coffee farmers, banana plantations represented the initial example of agribusiness, the application of industrial methodologies to farming on a large scale. The UFCO was a pioneer in this approach, at least in tropical settings, and the challenges it encountered forced it to seek new technologies that would support its position on the cutting edge of the industry. The economies of scale achieved on UFCO plantations generated the capital needed to fund a research capacity unprecedented in the annals of tropical agriculture.

The empire's built environment was also substantial. The UFCO's plantations, impressive for their facilities in areas where other infrastructure was generally lacking, provided excellent examples of the application of agribusiness principles to banana farming. Collectively, they created a legacy in Latin American banana production that growers in other regions found difficult to emulate. A typical plantation featured miles of irrigation ditches and overhead

3. Hauling bananas to the packing shed in Costa Rica.

4. Measuring and classifying bananas before packing in Honduras.
5. Packing bananas in Honduras.

pulley systems installed to carry fruit from banana plant to packing shed with a minimum of bruising. Packing sheds also underwent alterations over the years, as new technologies changed the nature of post-harvest handling.

The remote locations required that company housing be provided on many banana plantations. Each section of constructed housing reflected the rank of its occupants. At the top end were comfortable bungalow-style homes for the highest-ranking officials, generally expatriates. These represented the North American ideal of tropical comfort and today are still visible in rows along the waterfront in former company ports like Golfito, Costa Rica. More modest but usually well-built houses were provided for midlevel employees, a mixture of expatriates and nationals of the host country. Most

6. Former United Fruit Company plantation manager's housing in Honduras.
7. Plantation workers' housing in Costa Rica.

common were the rough-board longhouses for plantation laborers. Often these were built on stilts and were situated around a soccer field. They are still commonly seen throughout the banana zones of Central America. Although Kepner and Soothill (1935, 28–29) described such housing as "decidedly overcrowded," it exceeded the standards of many surrounding communities.

Special facilities were also present, including the commissaries noted earlier. The UFCO constructed eleven hospitals, one in each of its geographic divisions. It ran twenty-four radio stations, providing a communications link between North and South (Kepner 1936, 18). Other facilities included schools, electric plants, sewage systems, office buildings, ice plants, and recreational facilities such as golf courses and swimming pools for its expatriate staff.

The Empire at Its Peak

By 1930 a very high concentration of the global banana industry rested in just a few hands. The UFCO alone handled 65 million of the 103 million stems of bananas traded internationally that year, with most of the remaining 38 million controlled by Standard Fruit and a few other corporations. The UFCO's assets, including land, were impressive. Of its total landholdings, just 14 percent (450,000 acres) was improved in some manner, with only 189,000 acres planted in bananas at any one point in time. The company also had extensive landholdings in sugar cane and cocoa cultivation. The remaining 86 percent was unused, with potentially productive land held aside for later development as needed. That need often resulted from the environmental toll of banana cultivation. Soil exhaustion was rapid. Most land was productive for just five to ten years, after which it was abandoned (Kepner and Soothill 1935, 26–27). In such situations, the regeneration process often lasts thirty years or more.

Early writings about the banana industry often were uncritical. Emphasis was placed on the technological developments achieved by the UFCO and its rivals, and on the infrastructure that changed the face of much of the coastal tropics in Central America, Jamaica, and Colombia. Writing during the conservative 1920s, Reynolds (1927, 70) typified the evaluations of the industry's first stage:

> The Atlantic Coast of Central America offers ideal conditions for banana cultivation. It is in this region extending a few miles back from the Caribbean Sea, at an elevation not more than 250 feet above sea level—where there are hot days, humid nights, and an annual rainfall of from 80–200 inches—that the tropical jungle has given way to the greatest banana farms of the world. Within the past forty years an enormous agricultural industry, with its related interests of railways, docks, villages, stores, and hospitals, has sprung up in a region formerly almost uninhabited. Central America may well thank the banana trade for the most progressive influence and constructive development which have ever reached its shores.

It is easy to understand why early writers like Reynolds were so impressed with an industry clearly on the cutting edge of technology. Imperialism, whether formal or informal, still reigned in much of the South at that time and was viewed by some as the natural order of things. Most of Africa, Asia, and the Caribbean remained under European colonial rule. The neocolonial aspects of the Banana Empire would not have appeared unusual to many observers of that era.

This perspective began to change during the 1930s, when the first major critiques of the industry were published. Kepner and Soothill's *The Banana Empire: A Case Study in Economic Imperialism* in 1935, followed by Kepner's 1936 work, *Social Aspects of the Banana Industry*, presented a very different view of the industry and of the United Fruit Company. They analyzed the pattern in which transnational corporations with great amounts of investment capital sought opportunities in foreign lands where they set up large-scale operations. Once there, the companies purchased political influence to accomplish their goals. Kepner and Soothill (1935, 341) describe the essence of the resulting banana republic phenomenon:

The Banana Empire is not primarily an aggregation of mutu-
ally interacting governmental and industrial agencies, but the
expansion of an economic unit to such size and power that in
itself it assumes many of the prerogatives and functions usually
assumed by political states. The United Fruit Co., rather than
rely upon the State Department to pull diplomatic wires, trains
its own political representatives to deal with Caribbean governments. The corporation . . . not only monopolizes the banana
trade in the most important producing regions but also . . . can
speak with such force that politicians accede to its will.

On the occasions when local cooperation was not forthcoming,
UFCO did not hesitate to turn to its home government with requests
to intervene on its behalf. It was well connected to the sources of
U.S. foreign policy development, a legacy that continues today. U.S.
pressure usually was sufficient to gain local compliance with company demands. Where pressure failed, more direct action was taken,
as occurred later with the Central Intelligence Agency's (CIA's) well-
documented assistance in the 1954 coup that overthrew the democratic government of Guatemala. Kepner and Soothill's works analyze the UFCO's political clout during the first three decades of the
twentieth century, but the pattern of U.S. pressure they describe
does not differ significantly from the 1996 U.S. complaint to the WTO,
a complaint that European Union officials publicly state was originally written by the UFCO's successor company, Chiquita.

The empire thrived on Central America's lack of unity, its poverty,
and the weakness and easy corruptibility of its governments. It took
advantage of remoteness there and in poorly integrated, although
larger and stronger, Colombia to dominate the regions in which it
operated. Where such characteristics did not prevail, the UFCO did
not flourish. Jamaica and Mexico are two examples of countries in
which the company's efforts to control events related to the banana
industry did not succeed, suggesting that different outcomes would
have been possible in Central America had the above conditions not
prevailed.

The situation in Jamaica is described in chapter 4. In Mexico the major factors affecting the UFCO's plans were the Mexican Revolution, the relative unimportance of bananas to the national economy, and the existence of direct overland trading routes to the United States. UFCO was frustrated by the revolutionary context of events in Mexico after 1917, which favored the distribution of land to the peasantry and worked against the kind of land concentration preferred by the company. In addition, Mexico's economy was much more diverse; it was not dependent on bananas, which represented just 2 percent of its exports. The existence of overland connections to U.S. markets meant that the UFCO could not engage in its usual pattern of controlling transportation. Mexican banana farmers had other transportation options available to them (Kepner and Soothill 1935, 301).

Despite the unfavorable conditions, the company attempted to establish monopoly control over Mexico's banana industry. As production there increased during the early 1920s, making Mexico a competitor to the UFCO's Central American exports, the company began to purchase several export operators in Mexico's banana industry. It acquired additional properties during its takeover of the Cuyamel Fruit Company in 1929, leaving only Standard Fruit's subsidiaries beyond company control. Ultimately, the formation of a banana cooperative society in 1928, as part of a larger cooperative movement in post-revolutionary Mexico's rural sector, proved to be the determining factor in the United's failure there. Both the UFCO and Standard Fruit tried to thwart this development by working with individual planters, rather than with the association. They made loans to many planters to gain their cooperation, creating a debt bondage situation. Such activities often serve to undermine the successful operation of a cooperative society, but in Mexico the society was able to ship its members' produce to the United States and market it there under more advantageous terms than the UFCO was willing to offer. The company finally realized that the other options available to planters meant that monopoly control would not be achieved in Mexico. It ceded the rights of its Mexican subsidiary to Standard Fruit and withdrew from the country.

The exceptional situation in Mexico illustrates the true nature of the Banana Empire. The UFCO did not wish to operate in an arena where it had little hope of establishing monopoly control. Overall, that control would register several downsides, including the displacement of small-scale national producers and the elimination of competition by various surreptitious means. Its practices included the abandonment of communities during bust periods and the use of unfair pricing and schedule manipulation to control transportation systems. Finally, while paying handsome dividends to its U.S. shareholders, the company provided minimal contributions to national coffers as a result of the concessions offered in its contractual arrangements with host countries (Ellis 1983, 73–74). In the eyes of its critics, the costs of the company's control more than offset the technological benefits it offered.

The downturn in the global economy stimulated by the 1929 stock market collapse affected the functioning of the Banana Empire during the 1930s. This initiated a long process of change and moved the industry into a new stage in its development.

2

The Empire Challenged,
1930–74

By 1930 the essential ingredients of the Banana Empire were in place and the United Fruit Company operated with relative impunity throughout the so-called banana republics of Latin America. The Great Depression of the 1930s, however, was a major disruption to the banana trade and to national economies globally. It is often credited with inducing a new stage in the overall development of Latin America—the stage of economic nationalism—as national governments began to take a more active role in directing their economies. Ultimately, these changes would affect the nature of the Banana Empire.

This chapter covers the second stage of the banana industry's development. The origins of the second stage lie in the dire economic conditions of the Great Depression, but many of the effects of the Depression would not really be felt until the 1950s. Similarly, the roots of the third stage can be found in the growth of Latin American state sectors that began after World War II, a process initially manifested in the larger, stronger states of the region. For the smaller Central American states, Ecuador, and Colombia, the process began later and the strength needed to confront the Banana Empire was not mustered until the 1970s.

As the second stage of the banana industry's evolution began in the Western Hemisphere, the Banana Empire was at its virtual peak. The UFCO had created an unparalleled infrastructure in the countries in which it operated. The industry was modern and technologically advanced. Bananas were the leading export of Costa Rica, Honduras, and Panama and ranked highly among the exports of Guatemala, Nicaragua, and Colombia. Thousands of people had jobs directly related to banana production and trade.

But what was the state of the industry in 1930 from the perspective of the host countries? The facilities developed within the Empire were not intended to serve the societies at large. The Empire's infrastructure was focused on isolated banana zones and had weak, if any, connections to the heartlands of the countries involved. In Honduras the railroad system did not even reach Tegucigalpa. The industry was a classic enclave under foreign domination, with few links to the rest of the national economy in each producing country. Corruption of government officials ensured that states did not intervene in ways that would steer the industry's real benefits toward the host countries. Labor unrest was frequent but was commonly suppressed.

The industry also generated a relatively low volume of tax revenues for the host countries. Unlike developed countries, where personal income and property taxes generate a significant percentage of public-sector revenues, less-developed countries (LDCs) historically have a small middle class and thus are forced to rely on export and import taxes as a major source of funds. In Central America and Colombia, coffee and bananas were the two primary exports, but they did not share equally in providing funds for the national treasury. Coffee production occurred on modest-sized farms that were usually owned by nationals. Coffee exports were more effectively taxed than were bananas produced by large foreign corporations that negotiated favorable contracts with the government. According to Kepner and Soothill (1935, 212), export taxes paid to Honduras, Costa Rica, Panama, Guatemala, and Colombia on a nine-hand bunch of bananas having a wholesale value of two dollars or more ranged from zero to just one and one-half cents.[1] They estimated that, in Costa Rica, export taxes on coffee represented 11.8 percent of the export value while those on bananas were just 1.4 percent of the export value. Their analysis concluded that the governments of those countries "relinquished their sovereign right of taxing one of their two leading industries for many years" (Kepner and Soothill 1935, 212).

During this second stage, the Banana Empire encountered several challenges to its hegemony. The first challenge—Panama Disease—resulted in the company's developing the kind of geographic flexibility that has come to symbolize the footloose nature of modern TNCs. Panama Disease, a fungus, was so-named because it first struck the UFCO's plantations in Panama's Bocas del Toro province during the mid-1920s. The disease spread rapidly to the Caribbean zones in other Central American countries during the 1930s, wiping out the output of entire plantations—indeed, of entire production zones. Losses were serious and required creative solutions, leading to the first major geographic shift in banana production in the region. The company began to purchase more land on the Pacific side of the isthmus—primarily in Guatemala, Costa Rica, and Panama—which had not been affected by the disease. It developed new banana ports at Golfito and Quepos in Costa Rica and Puerto Armuelles in Panama to serve those plantations, and it increased its use of existing Pacific ports elsewhere. The abandoned Caribbean banana zones were devastated, the vulnerability of their specialized economies clearly exposed. The states involved were still too weak to transact meaningful interventions. The UFCO maintained most of its real estate holdings on the Caribbean side to keep open the option of a future return, which did eventually occur.

A geographic shift also occurred at a later date and for different reasons in Colombia, away from the Santa Marta region. Continuing labor problems and the Depression-induced crisis led the UFCO to reduce its investments in the area (Sierra 1986, 2–3). By the 1960s, due to declining competitiveness, the UFCO moved most of its Colombian operations from Santa Marta to the Urubá region; the company began to export bananas through the Gulf of Urubá in 1962 (Sierra 1986, 8).[2] Such moves had a serious impact on the region left behind, including loss of employment. The jobs lost were usually higher paying than those outside the banana industry; often they were among the best-paid jobs in the agricultural sector. The departure of banana jobs carried the added burden of the loss of housing for many of the released employees, contributing to housing deficits in the affected regions.

The Great Depression provided a second great challenge to the Banana Empire. It reduced demand in consumer countries, lowering the UFCO's earnings in the process and causing the company to become even less willing to deal fairly with workers and governments. But the Depression, together with Panama Disease, stimulated the company to explore new strategies that would lead to substantial change in the industry. Most of the changes had to wait, however, while a third major challenge to the functioning of the empire— World War II—ran its course. During the war, commercial vessels, including those of the Great White Fleet, were pressed into public service. Banana exports from the four major Central American producers continued during the war, but on a reduced scale, sinking to just 12.3 million stems in 1943, their lowest level since 1910. The return of the fleet in 1945–46 resulted in a rapid recovery, with exports rising to 35,869,349 stems and more than 53 percent of the world's total banana exports in 1946 (Ellis 1983, 54).

With the industry recovered, the importance of bananas to national economies once again became evident. In 1947 the percentage of gross national product (GNP) generated by bananas among the four major Central American producing states ranged from a high of 38.7 percent in Honduras to a low of 12.3 percent in Panama, where the

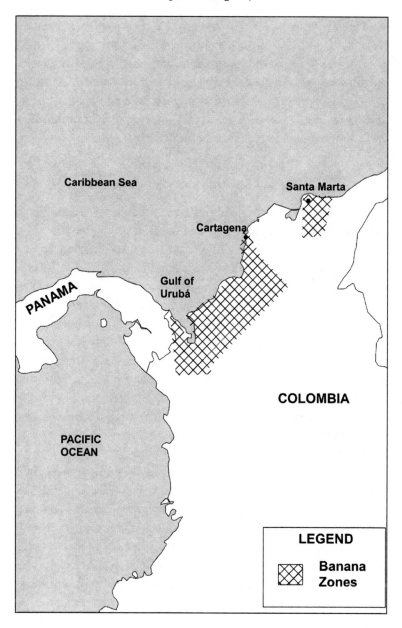

3. The banana zones of Colombia

service sector dominated the economy. Costa Rica (16.5 percent) and Guatemala (22.7 percent) fell in between these extremes (Ellis 1983, 71). Bananas figured even more significantly in the trade profiles of the four countries, representing 78 percent of the export earnings in Honduras, 72 percent in Panama, 46 percent in Costa Rica, and 42 percent in Guatemala (Ellis 1983, 72). Coffee exports from the latter two countries explain the lower impact of bananas there.

Ecuador Becomes a Major Player

Although bananas had been cultivated in Ecuador for centuries, the country was not a major exporter of the fruit during the first stage of the banana industry in the Americas. That situation changed rapidly with rising demand in the years immediately after World War II. Along with a better transportation infrastructure and the availability of cheap agricultural credit, the increased demand, especially in Europe, fueled a great expansion in banana production in the country's coastal regions. By 1952 Ecuador occupied first place among the world's banana-exporting nations. As other Latin American exporters began to shift some of their exports from U.S. to European markets, Ecuador's fruit filled the gap in the United States. Virtually all of its banana exports to the United States passed through the Panama Canal to Gulf Coast and East Coast ports, rather than to the West Coast.

Ecuador had several advantages in this new endeavor. Because it had not previously experienced large-scale banana production, its territory initially was relatively free of the plant diseases afflicting bananas elsewhere. Its location on the equator protected it from the hurricanes and other tropical storms that occasionally devastated Caribbean and Central American production. The climate also offered ideal growing conditions for the fruit, with sufficient moisture and productive soils.

The banana zones that emerged in Ecuador were concentrated in the western third of the country. The Guayas lowlands, in

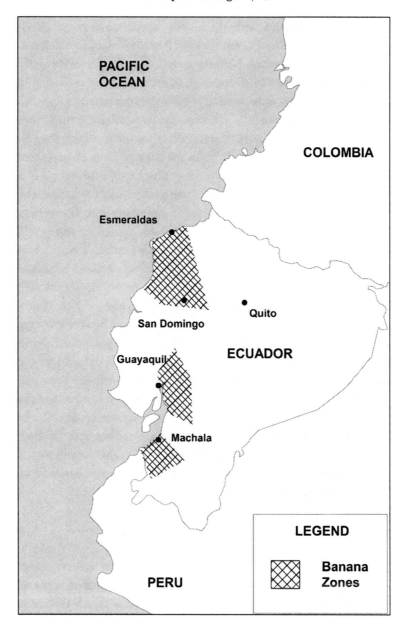

4. The banana zones of Ecuador

Guayaquil's hinterland, were the first to experience large-scale conversion of tropical forest to plantations. It was followed in quick succession by the Esmeraldas hinterland in the north, a corridor along the new road linking Santo Domingo to the port at Esmeraldas, and the southern coastal zone around Machala (Parsons 1957, 201).

In the late 1940s Ecuador initiated a homestead program to lure farmers from the Andean highlands, which had always been the cultural heartland of the country. Titles for 120 acre plots of land were offered to farmers on the condition that they begin cultivation on at least one-fourth of the land within a five-year period. Because of the relatively brief period (nine months) from planting to harvesting, bananas were the ideal crop for meeting this requirement and would generate income more rapidly. The new roads permitted easy access to ports for the export of the fruit. The initial result was a model of relatively small-scale production, with most farms having just ten to thirty acres planted in bananas (Parsons 1957, 203–4). A small percentage of fruit was grown on plantations owned by foreign TNCs, especially the UFCO, but the UFCO did not rise to the position of dominance that it occupied in Central America. Railroads did not play the same role in Ecuador as in the countries to the north; government-built roads, funded through loans from the U.S. government and the World Bank, provided the necessary means of transportation. While the UFCO and Standard Fruit did control much of the international shipping stage of Ecuador's banana industries, the UFCO sourced two-thirds of its bananas and the Standard at first purchased all of its bananas from national growers.

Labor Issues

Labor was the empire's fourth challenge during the industry's second stage. The 1930s were a period of great labor activism around the world, and the banana industry reflected this trend. Efforts at unionization were common. The UFCO resisted such activism with great vigor and displayed considerable political influence in the banana-producing countries. With government support, often in the form of troops, the company used strikebreakers to thwart labor's attempts to improve the situation for workers.[3]

The divided nature of the labor force enabled the UFCO to exercise greater control. During the first stage of the industry, most workers were of Afro-Caribbean origin; many were English speakers and were culturally different from the people of the highland core regions of Central America. Spanish-speaking Latinos often found the disease-ridden tropical environments of the banana zones repulsive. Their attitudes changed over time, as the company invested substantial effort in researching tropical medicine and improving sanitation conditions, endeavors that were funded through 2 percent deductions from employees' pay. The improvements attracted greater numbers of Latinos into the banana zones, increasing competition for jobs and driving wages downward. The competition increased ethnic divisions within the workforce, which the UFCO was quick to exploit, making it more difficult to achieve labor solidarity in confronting the company (Kepner and Soothill 1935, 317). With its vertically integrated system, the UFCO controlled most other employment opportunities in the banana zones and port areas, further limiting the options available to workers.

Where possible, especially in the initial stage of clearing rainforests for new plantations, the company used contractors to assemble a labor force. In such cases, the UFCO and its associate producers had no direct responsibility for the workers, nor for social costs such as workmen's compensation and insurance. This reflects a now-familiar theme. In 1994 Costa Rica's press published several articles about subcontracted labor in the banana industry and the lack of social payments on their behalf (*La Nación* 1994a and 1994b; *Tico Times* 1993). The issue arose within the wider debate over the use of Nicaraguan workers in the country's agricultural sector.

The retrenchment stimulated by the Great Depression fomented greater labor unrest and prolonged strikes that often became violent. Increasingly difficult working circumstances provided ripe conditions for the region's communist parties, which became involved in the workers' cause. Unfortunately, their influence over the newly forming unions often served as a lightning rod for those more opposed to communism than to unionization generically. This provoked an even stronger reaction to labor activism than might otherwise have been the case. In Costa Rica, for example, the Communist Party assisted workers in founding the Sindicato de Trabajadores del Atlántico (Workers' Union of the Atlantic) in 1934. A violent strike followed, generating stories that achieved legendary status in that country.[4] The violence cut both ways; strikers destroyed fruit and uprooted railroad tracks and switches to prevent the export of bananas harvested by strikebreakers (Kepner and Soothill 1935, 331). Predictably, the country's traditionally conservative press branded the strike a communist menace.

Eventually, as populist governments rose to power throughout Latin America, unionization progressed and workers' rights were enhanced. Social legislation also improved, although remaining far below standards in developed countries. A notable exception was Costa Rica, where significant advances were realized during the administration of Calderón Guardia (1940–44). According to Antonio Montero (1994, personal communication), director of the Instituto Centroamericano de Asesoría Laboral (Central American Institute for Labor Advisement), during the 1940s, Costa Rica was fortunate to have a president, labor union leader (Manuel Mora), and church leadership who shared similar progressive views, creating the synergy needed to break with the past. Together they laid the foundations of the country's widely hailed social security network.

Continuing communist involvement with the country's unions, however, caused a conservative reaction that led to the creation of an alternative form of labor organization that was less threatening to the established order. The formation of *solidarismo* organizations began in Costa Rica during the late 1940s and spread to other countries of the region. Solidarismo was modeled after the labor methods of Fascist Italy and Germany. These organizations linked company owners and workers together in a way that essentially co-opted the latter. The company's principal goal was to avoid conflicts (i.e., strikes) while limiting concessions to the workers. One of the primary features of solidarismo organizations is mutual or workers' funds. These funds, which today function more like pension plans, were originally intended to give workers a stake in a company's prosperity.

Solidarismo did not progress very rapidly in Costa Rica through the 1960s. It was mostly found in companies owned by nationals and did not yet operate in the banana zones, where leftist unions were still strong and the government had relatively little power vis-à-vis the foreign-owned fruit TNCs.[5] Solidarismo did get a boost from the new *maquiladora* (foreign-owned assembly plant) sector in the 1970s, but its penetration into the banana zones would not occur until after the banana had entered its third stage.

Panama Disease Strikes Again

The dreaded Panama Disease continued to haunt the UFCO and its rivals, rising again during the 1950s to challenge the Banana Empire and causing great financial losses to the fruit companies. The situation was so serious that the United decided to restructure its divisions to save money. It streamlined its Central American operations and ran them directly from Boston, allowing greater coordination of its activities. On occasion the UFCO had to purchase bananas from other sources, particularly Ecuador, to replace the bananas destroyed on its infected plantations. The company even considered a plan to abandon its Central American plantations and rely on Ecuador and the Philippines to supply its markets, but its huge investment in the region ultimately deterred such a decision. Most of the UFCO's assets were tied up in land and infrastructure—much of it not moveable—in the Central American republics.

Instead, the UFCO sought to resolve the problem of Panama Disease by abandoning the disease-prone 'Gros Michel' banana and searching for a resistant variety. Standard Fruit, facing the same situation on a smaller scale, turned to the 'Giant Cavendish' variety in the late 1950s, but it proved too fragile and easily damaged in storms and during shipping. The UFCO adopted another Cavendish variety, the 'Valery Cavendish,' which it introduced in 1960. The 'Valery' was disease resistant and offered the additional advantage of being a shorter plant that was less likely to be damaged by winds (CEPAL 1979, 61).

The change from the 'Gros Michel' to a Cavendish variety instigated a basic change in the post-harvest handling of bananas. Previously, bananas were shipped on the stem and boxed in the ports of entry in the North. As the Cavendish bananas are easily bruised, they must be handled in a more delicate manner. The new process involved de-stemming and boxing on site for the first time on a significant scale. Ironically, this allowed the UFCO and its rivals to circumvent a 1957 U.S. Court antimonopoly ruling that prevented large fruit firms from engaging in those activities within the United States, offering an early example of moving an operation offshore and taking advantage of the lower labor costs in the producing countries. Boxing on site also permitted the companies to reject poor-quality bananas before they were shipped, reducing the weight of their cargoes and their total shipping costs. Finally, the new system also generated an ancillary activity that represented a new profit center for the vertically integrated industry—the manufacture of the standardized 18.14 kilogram (40 pound) cartons.

The Panama Disease crisis also precipitated another major change in the nature of the Central American banana industry—a partial, although significant, retreat by the UFCO and its rivals from the banana cultivation stage. This stage is fraught with risk and generates just a small percentage of the industry's total profits. The benefits of reducing exposure to risk by farming out some of the production became obvious, although doing so would require a system that the companies could control and manipulate. As Panama Disease again grew to a crisis during the 1950s, the companies began to put such a system in place for an increasing percentage of the fruit they shipped northward.

The Rise of the Associate Producers

The large banana companies had always purchased a portion of the fruit they eventually marketed in the United States and elsewhere. This was the case during the first Panama Disease epidemic and during episodes of soil exhaustion (Kepner 1936, 51). The suppliers were nationals of the exporting country. During the first stage of the industry, these so-called independent producers farmed their own lands and entered into individual contracts with the dominant fruit company in their region. These farmers were small-to medium-scale entrepreneurs, but they lacked the technologies being developed by the TNCS, particularly by the UFCO, which had created a research division in Central America. True independence was not attainable for the growers, as they were dependent on the TNCS for all of the forward linkages in the industry, including transportation and marketing of their fruit. They also had no influence over the prices paid for their bananas, and their contracts tended to be long-term relationships with little flexibility built into the pricing system.

During the 1950s this system was formalized into associate producer programs in each of the major Central American exporting countries, the first of which was established by the UFCO in Honduras in 1954. It served as the model for such programs through the early 1970s (Paredes 1976, 8). Many associates were former employees of the company who now leased its lands. Emphasis was placed on technology transfer, as the companies provided technical services and training to yield higher-quality fruit. Financing was also available, and the programs included access to superior grade plant stocks, fertilizers, and increasingly, agrochemicals used in pest control (OAS 1975, 15). None of this was free, of course. Growers paid for these inputs and services. Also, it should be noted that the existence of associate producer programs does not imply the existence of associations of producers. Associations did not enter the picture until the early years of the industry's third stage.

After 1960, changing conditions in the industry stimulated development of a more refined model of the Honduras program. The associates acquired greater responsibility for investment, production, packing, relations with unions, and all socioeconomic aspects of employee relations. Their contractual relations with the TNCs specified fixed prices, guaranteed purchase of fruit under quality criteria, and technical assistance. The benefit of such arrangements to the growers included the opportunity to develop as business operators and to diversify their operations if they chose to do so (Paredes 1976, 9).

The growth of the associate producers programs continued after the Panama Disease crisis was resolved. The TNCs recognized the benefits they received from such relationships. Foremost among these was the transfer of risk from company to grower for an increasing percentage of the fruit marketed. Also important was the buffer provided by such sourcing. When prices or demand declined, the company could employ more stringent quality standards and reject higher percentages of the growers' fruit. Although quality standards were built into the contracts between company and grower, their interpretation was subjective, allowing flexibility for the company's representative at the point of transaction. Another benefit involved labor; many of the companies' labor problems were transferred to the associate producers, who functioned as the direct employers of a growing number of workers. This benefit increased over time, through the 1960s and early 1970s, during periods of labor activism and rising nationalism in the region. The banana TNCs pioneered the establishment of the associate programs, but the system was later adopted for other commodities as well.

Overall, the shift from company to associate producer was substantial. In the case of Standard Fruit, the share grown by its associate producers increased from 26 percent in 1968 to 42 percent in 1974. During that period, the company's overall shipments of bananas from Central America increased by more than 37 percent (OAS 1975,

18). For the UFCO, production declined in the same period, from 77 percent to 62 percent of total marketed fruit, with the balance purchased from associate producers and from sources in Ecuador and the Philippines (OAS 1975, 20).

The system functioned well for both sides until 1970. The associate producers enjoyed good earnings and the TNCs reaped benefits from the relationship. After 1970, however, many growers encountered financial difficulties in their non-banana activities. They found the companies inflexible in negotiating new contracts, which increased fiscal stress for many growers (Paredes 1976, 10). Coupled with the supportive political climate of the early 1970s, the financial stress encouraged many associate producers to resist the control of the TNCs. They sought both increased state involvement in the industry and the opportunity to create growers' associations without company approval. This trend contributed to the so-called banana war of 1974, which ushered in the industry's third stage.

The Field Grows More Crowded

UFCO's competitors provided the sixth challenge of the empire's second stage. During the banana industry's first stage, the company had eliminated or contained its rivals through takeovers and exclusionary contracts with national governments, preventing other fruit companies from gaining footholds in its regions. Only Standard Fruit had a substantial presence in Central America, and its operations did not overlap with UFCO's territory. As the 1950s arrived, the company still controlled 80–85 percent of the U.S. banana market, the world's largest market at that time. Its vertical integration strategy enabled it to reap profits from each stage of the industry's operations, up to the point of retail sales.

This situation changed during the 1950s—ironically, during the Eisenhower administration, when officials of the administration aided the Banana Empire in achieving what might be considered its greatest victory: toppling Guatemala's democratic government. The CIA-assisted military coup that ended Guatemala's ten-year "springtime of democracy" in 1954 illustrated the UFCO's far-reaching political influence. Its governmental connections included Secretary of State John Foster Dulles, who was a former UFCO lawyer, and his brother, the CIA director and a member of the UFCO board. The Jacobo Arbenz government in Guatemala expropriated four hundred thousand acres of uncultivated UFCO land, prompting the company to capitalize on the high level of anticommunist fervor in the United States by seeking U.S. intervention in Guatemala's internal affairs. The subsequent coup overthrew Guatemala's democratic government, replaced it with Colonel Carlos Castillo Armas, a friend of the CIA and the UFCO, and plunged the country into a period of repression, instability, and civil war that lasted until 1998 (Dosal 1993, 229).

But it was also during the Eisenhower administration that the UFCO encountered a new challenge to the way in which it conducted business. Antimonopoly activity by the U.S. Department of Justice forced the company to sign a 1958 consent decree rather than face protracted legal proceedings. The terms of the decree included divestiture of some of its landholdings and infrastructure in the United States and Central America. Domestically, the decree prohibited the company from holding title to bananas beyond the port of entry. It would now have to yield title to ripeners or chain stores at the point that the bananas were transferred to refrigerated cars, allowing other companies to earn the profits from ripening and distributing the fruit (OAS 1975, 26). Abroad, the decree provided space for competitor firms and led to the emergence of a triumvirate of U.S. fruit TNCs, with Del Monte and Standard Fruit taking their places alongside the UFCO.

The Del Monte Corporation is the world's largest canner of fruits and vegetables and a major supplier of canned seafood and dried fruits. During the 1960s the corporation wanted to diversify its operations. In 1968 it purchased a small Costa Rican firm, the West Indies Fruit Company, renamed it the Banana Development Corporation (BANDECO), and entered the banana industry. This was a small challenge to the UFCO's operations in Costa Rica, but Del Monte's real motivation was to prevent UFCO from making a takeover bid. The UFCO had acquired seven hundred thousand shares of Del Monte stock by 1968, 6 percent of the company's total (CEPAL 1979, 102). But because U.S. antitrust laws prohibit a company from purchasing a direct competitor, the UFCO could not proceed with its intended buyout of Del Monte.

With a foothold thus secured, Del Monte moved to expand its banana operations. The aforementioned consent decree worked to Del Monte's benefit as it required the United to divest itself of its Guatemala landholdings. The United needed a buyer, and Del Monte was interested. They began to negotiate a sale, but the Guatemalan government objected, preferring a Guatemalan owner for UFCO's subsidiaries in the country. This delayed negotiations, but they resumed after Del Monte contracted a Cuban-born Guatemalan entrepreneur, Domingo Moreira, to serve as its consultant. According to Burbach and Flynn (1980, 209), Del Monte surreptitiously channeled five hundred thousand dollars in funds from its Panamanian subsidiaries through Moreira to strategic Guatemalan officials who subsequently permitted the sale to proceed. Thus, in 1972 Del Monte bought fifty-five thousand acres of UFCO property and infrastructure in Guatemala for twenty million dollars, doubling its Latin American holdings overnight (OAS 1975, 44).[6]

Despite its late entry into the field, Del Monte quickly emerged as a major player. Its banana division, the Del Monte Banana Company, became the third largest marketer of bananas in the United States. It owned plantations in Guatemala and Costa Rica. Its Guatemalan subsidiary, the Banana Development Corporation of Guatemala (BANDEGUA), basically picked up where the United left off. It used the same strategies with labor and with the government, to whom it paid no taxes on its land. Forty-five thousand acres that were not planted in bananas were used for grazing seven thousand head of Del Monte–owned cattle—not for the marketing of beef but simply to deter squatting, which was legal on unused private land in Guatemala and elsewhere in Latin America (Burbach and Flynn 1980,

211). With its workforce, the company continued the United's practice of paying relatively high wages by agricultural standards and provided a benefit package that often included housing and electricity. But it co-opted the leadership of the workers' union, associating the organization with the U.S. Agency for International Development–supported American Institute for Free Labor Development, which was, according to Burbach and Flynn, "used by the CIA to undercut genuine progressive unionism in Latin America" (Burbach and Flynn 1980, 215).

In the meantime, the United itself underwent significant change. In an effort to diversify its operations and improve its image, the company merged with the AMK Corporation in 1970 (CEPAL 1979, 64). This move was accompanied by a name change to United Brands (hereafter, UB). The resulting company had a more diverse portfolio that included meatpacking and food processing (CEPAL 1979, 58). The UB selected a new chief executive officer, Eli Black, whose management skills put the company's fiscal house in order following the second Panama Disease crisis. Black's aggressive stance against competitors, however, unintentionally led the industry into its third stage. He launched a banana price war in 1970. But the price war backfired, and by 1974 UB was no longer the leading supplier of bananas to the U.S. market (table 1). Standard Fruit overtook it following UB's sale of its Guatemalan holdings. But the company still had a diverse market profile, dominating the market sales to Canada, the UK, and several Western European markets.

The Standard Fruit Company also was active during this period. It was purchased by Castle and Cook Inc., a diversified food company, and became its banana subsidiary in 1968. Standard Fruit had substantial assets, including land, in Latin America. It owned 163,000 acres in Honduras and had smaller holdings in Costa Rica and Nicaragua (CEPAL 1979, 85). It also had subsidiaries engaged in activities like pineapple cultivation, cattle grazing, and railroads. Castle and Cook also assumed control of Standard Fruit's associate producer programs, including those in Ecuador. Finally, it carried on the Standard's role as a major purchaser of bananas in the Philippines.

TABLE I. U.S. market share for major banana importers (%)

Year	UFCO/UB	Standard	Del Monte	Others
1950	80	8.9	—	11.1
1973	34.6	40.8	15.6	7.2

Source: OAS 1975, 24

During this period, Standard Fruit was responsible for a major marketing innovation in the industry. Castle and Cook was known primarily for producing pineapples in Hawaii and marketing them under the Dole brand name. Bananas had not previously been marketed by brand name, but Standard introduced the concept during the 1960s, using the Dole name to engender consumer loyalty and improve its market share. Eventually, UB followed suit, adopting the now familiar Chiquita brand name in 1968 (Wiley 1994, 67).

As the banana industry's second stage drew to a close, the impact of the consent decree was quite clear. Table I indicates the shift that occurred after 1950. Not only did UB lose its near-monopoly control in the U.S. market, it also lost its position as the primary supplier to Standard Fruit. By 1973 the United States, which was still the world's largest market and imported 27 percent of all bananas traded internationally, had a market that was shared among the Big 3 banana TNCs. Together they controlled more than 90 percent of all U.S. banana imports (OAS 1975, 21).

By 1973 all three banana companies were part of larger, more diversified food companies. These horizontally integrated corporations were able to use earnings in other operations to offset annual losses in their high-risk banana divisions, and banana profits could return the favor in good years. In 1973 the companies and their subsidiaries handled 53 percent of the world's banana exports, including two-thirds of those from Latin America, and 55 percent of the world's banana imports. Within one year, however, all three encountered a new challenge that led the industry into a third stage.

3

The End of Splendid
Isolation, 1974–93

By 1974 Latin America had been engaged for several decades in a process of economic nationalism, but the length of this process varied among individual countries. Import substitution industrialization (ISI) programs, which were intended to increase secondary sector activity while reducing costly manufactured imports, reshaped national economies throughout the region. ISI was implemented by national governments whose involvement in the economy grew relative to the involvement of private sectors. Under state capitalism, governments participated directly in the production process and owned tertiary sector firms such as banks, airlines, and telephone companies.

The motivation for these state-led development programs came from the Great Depression and the two world wars earlier in the century. Those crises disrupted Latin American development while exposing the region's dependence on developed countries in the North. These problems inspired new theoretical frameworks and methods of analysis for understanding development processes and global capitalism—dependency theory and structural Marxism among them. A new intellectual school emerged, represented within the United Nations Economic Commission for Latin America under Raul Prebisch. It challenged traditional approaches to development, focusing on the external causes of Latin America's underdevelopment, and influenced the ISI programs and other initiatives pursued by the region's governments.

The ISI industries that emerged encountered many problems. The small size of internal markets was particularly serious because it prevented most of the industries from deriving the economies of scale needed to become competitive. Many companies required protection from competition abroad and had to be subsidized from public-sector coffers. Over time they became expensive to maintain, causing governments throughout the region to seek new funding sources for strained treasuries. Most borrowed from abroad, leading to the 1980s foreign debt crisis that ravaged Latin America. Banana-exporting countries were among those seeking loans, but they also decided that the time had come to derive more revenues from one of their most important products.

The inspiration for this momentous decision might well have come from the Organization of Petroleum Exporting Countries (OPEC). The first oil crisis (1973), precipitated by events in the Middle East, increased the cost of Latin American countries' highly subsidized energy sectors, and the higher price of imported oil caused an additional drain on public-sector funds. At the same time, however, the OPEC strategy offered a new example of a group of developing countries, including Ecuador and Venezuela, challenging the established economic order to derive greater benefit from a primary sector export (Wiley 1994, 68). Several countries in the region believed that such an approach might work for bananas as well.

The Banana War of 1974

The 1990s battle over bananas between the United States and the EU was not the first banana war. Langley and Schoonover (1995) describe skirmishes during the first decades of the twentieth century between Central American governments, notably in Honduras and Nicaragua, and the North American banana entrepreneurs seeking to unseat them. In addition, Guatemala and Honduras nearly went to war over banana zones during the industry's second decade.

A more far-reaching struggle occurred in 1974 when the governments of several banana-exporting countries, particularly Costa Rica and Panama, began to reevaluate the role of their banana industries within the wider framework of national development. This significant step signaled a desire to change the historical enclave nature of the industry, whose principal benefits to the countries involved were the introduction of technology, expansion of the realm of productive lands, and remuneration packages for workers (OAS 1975, 3–4). Tax revenues remained low and the industry was not regulated or monitored by the state, except in Ecuador where most owners were nationals. The relationship between the countries and the companies was defined by long-term, binding contracts that lacked the flexibility necessary to respond to changing conditions.

Small steps had been taken to achieve international cooperation among governments of banana-exporting countries. Generally, these occurred within the frameworks of organizations like the Organization of American States (OAS), which became involved with banana issues during the 1960s and opened the Italian market to Latin American bananas in 1962. An International Banana Group was created within the United Nations Food and Agriculture Organization (FAO) in 1964. Neither of these groups, however, had yet confronted the real power centers of the industry—the large, transnational fruit companies.

That step was finally taken in 1974. It was precipitated by the price war among the three TNCs that began in 1970 when United Brands' Eli Black attempted and failed to drive out UB's competitors. But Standard and Del Monte did not yield. The associate producers suffered the most because they were at the mercy of the TNCs to market their fruit, and they tried for four years to negotiate a solution

(Gabrielli 1988, 6). Their situation finally attracted the attention of the region's politicians at a time when nationalistic fervor was rising to new heights throughout Latin America. The increased attention led to the 1974 banana war after leaders like José Figueres in Costa Rica and General Omar Torrijos Herrara in Panama, indignant about their countries' continuing images as banana republics, became concerned about the poor prices for such an important export. Figueres, a charismatic figure in Costa Rica, was revered by its people, who lovingly called him "Don Pepe." He led the country out of its 1948 civil war, abolished its army in 1949, and in 1974 was concluding a second nonconsecutive term as president. Torrijos became a symbol of Panamanian nationalism when he negotiated the 1978 treaty that ultimately provided for Panama's control over the canal in 2000. Gabrielli (1988, 5) suggests that Torrijos represented an "indisputable leader of a new philosophy and relation between the banana producing countries and the fruit TNCs." It is no accident that the first successful challenge to the foreign-owned banana companies by Latin American states occurred under such leadership.

Ironically, the "war" began at a time when the newly formed UB was already concerned about the image it inherited from the UFCO. The company hired a consulting firm to analyze the situation and recommend changes to improve its image in a region characterized by increasing nationalism and expropriations. The recommendations included renegotiation of UB's old contracts with host countries to create a more modern relationship; the transfer to the state of functions that the company had previously operated, such as schools, hospitals, and water supply; the sale of housing to its workers to break the company town image; and payment of taxes based on banana production levels, not on the use of utilities or facilities (Gabrielli 1988, 5). If adopted, the recommendations would have prompted an overhaul of how the company transacted business in Latin America, negating three-quarters of a century of past practices. However, UB initially rejected the consulting firm's ideas, although its stance softened after the banana war began.

Concern over the ongoing price war stimulated a series of consultations between Torrijos and Figueres in early February 1974. The Honduran president quickly joined the discussion, which focused on how to generate more income for the banana-producing countries. Later in the month, the consultations expanded to include Colombia, Ecuador, Nicaragua, and Guatemala. The leaders used phones, personal meetings, and emissaries to communicate. Although the TNCs did not cultivate bananas in Ecuador, Standard Fruit continued to transport and market Ecuadorian fruit. Ecuador's status as the leading exporter made its participation in the group desirable if a coordinated effort to pressure the TNCs was to succeed. Much of the leaders' rhetoric centered on solidarity and political will, and these were initially achieved.

The first public volley in the banana war was hurled in late February, when Figueres announced that the countries desired to increase export taxes. The TNCs, especially UB, immediately began to register their opposition and generate a disinformation campaign and other tactics to try to dissuade the governments from this path. Their efforts included threats to suspend exports due to the huge losses they would suffer if the tax were imposed.

On March 8, 1974, the ministers of Colombia, Costa Rica, Ecuador, Guatemala, Honduras, Nicaragua, and Panama signed the landmark Panama Accord. They committed themselves to act as a front to raise export taxes on bananas by a range of forty cents to a dollar per box and to create a cartel organization, the Unión de Países Exportadores del Banano (Union of Banana Exporting Countries), known by its Spanish acronym, UPEB.[1] On March 14 the TNCs reiterated their threat to suspend exports, and the participating countries responded by agreeing to punish any country that broke the chain of solidarity. The solidarity, however, did not last very long.

On April 1, Panama and Costa Rica implemented an additional tax of one dollar per box, Honduras an additional fifty cents per box, and Colombia an additional forty cents per box. Ecuador, Guatemala, and Nicaragua did not immediately impose an increase

(CEPAL 1979, 72).[2] UB decreased its harvest and exports, and the TNCS sought support among U.S. dockworkers, asking them not to move cargo going to or from a country that raised its tax. As a scare tactic, UB and Standard Fruit offered in June to sell their farms in Panama and Costa Rica, respectively, to the countries' governments. The two countries considered the companies' offer, and Costa Rica decided to call their bluff, announcing that it would buy Standard Fruit's holdings in Costa Rica. The next day the company paid its additional export taxes (Gabrielli 1988, 11).

Since 1974 the taxes imposed on banana exports have risen and fallen with changing conditions in the industry. Following storms that damaged banana plantations, the affected country generally reduced the tax to lower the financial burden on the companies while they repaired their farms. Costa Rica pursued a sliding scale tax policy through which it raised taxes when banana prices were high and lowered them when they fell. Overall, the new system of taxes resulted in considerably greater tax revenues for those countries that increased their export tax. Gabrielli (1988, 12) estimates that more than nine hundred million dollars in additional taxes were raised by 1988—modest by U.S. standards, but nevertheless a substantial flow into rather small national treasuries.

Institutionalizing the Banana

The Panama Accord of March 1974 marked an advance in the institutionalization of the banana. During the industry's first two stages, the banana TNCS functioned within the "splendid isolation" that they successfully carved out in their respective banana zones. Their activities were often shrouded in mystery. The host countries' governments knew relatively little about their operations and had only a rudimentary understanding of the industry itself. This situation changed substantially in 1974. Henceforth, the banana industry would be surrounded by a variety of state, parastatal, international, nongovernmental organizations (NGOS), and research organizations, each contributing to, regulating, or monitoring the industry to a degree that was previously unimaginable. Today the FAO and the OAS have a major presence, the latter through a subsidiary, the Inter-American Institute for Cooperation on Agriculture (IICA). These organizations, along with specialized schools focusing on tropical agriculture, are now on the cutting edge of research on issues related to the cultivation, marketing, and environmental aspects of the banana industry, issues that formerly were the preserve of the TNCS, especially the UFCO.

The creation of the UPEB in September 1974 was an important step in the process of institutionalization. The UPEB Convention was signed by the governments of Colombia, Costa Rica, Honduras, Guatemala, and Panama. Ecuador was notably absent, which signified a break in the Latin American unity achieved earlier that year. UPEB's OPEC-like goals were laid out in Article 2 of the Convention, including establishing and defending remunerative prices; promotion of common policies regarding the production, transportation, and marketing of bananas; working to expand into new markets; establishing an equilibrium between production and demand levels; cooperating in technical matters; modernizing the industry; and defending the participation of each member in international banana markets. Although the TNCS are not mentioned in the statement of goals, the activities noted represent a penetration into realms that the TNCS had traditionally controlled themselves.

The UPEB was not heavily bureaucratized. Its principal policy-making body was the Council of Ministers, composed of line ministers from the members. It was required to meet just once a year. More mundane policy matters were to be determined by the *consejo* (council), composed of representatives from each country. The UPEB Convention stipulated that each consejo representative had to be a national of his or her country, presumably to prevent foreign influence over the organization's decision making. Member states would provide the organization with financial resources raised from the export taxes on their bananas. Of the total budget, 25 percent was to be assessed in equal shares, while the remaining 75 percent would be assessed in proportion to each country's total banana exports. Voting power in the UPEB consejo was also based on the relative size of the industry in each of the participating countries (UPEB 1974).

One of the unstated goals of the UPEB was to provide a counterbalance to the power of the TNCs. To do this, the members needed to improve their knowledge of the industry and its operations. Panama's Fernando Manfredo spearheaded this effort, contributing to the UPEB's information base and fomenting the creation of formal institutions in the member countries. Each member authorized a parastatal agency to advance national interests within the industry.[3] Their task was to develop and implement what had never previously existed: a national banana policy. In most cases the new agencies invited participation by the TNCs, whose expertise in the industry was still valued but welcomed only within a framework where rules of corporate conduct could be applied. In recent years the parastatals have also become important sources of data on banana production and trade information.

Panama created its agency, the Oficina Nacional del Banano, within the Ministry of Trade and Industry in 1974. Its first annual report noted that its establishment was a direct result of the confrontation between the TNCs and the governments of the banana-exporting states (Oficina Nacional del Banano 1975, 1). It further stated that it was charged with the responsibility of studying and analyzing the industry so that Panama would no longer be ignorant of the conditions surrounding such an important activity and would be able to prevent "the exploitation of its natural resources by foreign hands" (Oficina Nacional del Banano 1975, 1). Such rhetoric reflects the nationalism sweeping Latin America at the time, and it yielded results. The Oficina Nacional used the new tax revenues to subsidize the country's associate producers to offset the low prices paid by the Chiriqui Land Company, a UB subsidiary, during the price war. Its research efforts contributed to the development of fairer contracts with UB. In 1975 Panama expropriated thirty thousand acres of uncultivated UB lands in Chiriqui province, and it acquired all of the company's holdings in the country within a four-year period (CEPAL 1979, 75–76).

There have been two UB contracts in Panama since the 1974 banana war, one in 1976 and the other in 1988 (Alegría 2000, personal communication). Under these contracts, the company cannot act to reduce or expand production areas, change the varieties of fruit cultivated, or introduce new technologies without consultation with the government. UB's obligations include production, exportation, good farm maintenance, payment of remunerative prices to its associated producers, and obedience to national laws (Gabrielli 1988, 13). Panama nationalized the land occupied by the company, but UB continues to own the infrastructure it placed there and pays rent for the land itself. After the negotiation of the 1976 contract, Fernando Manfredo commented: "In essence, the juridical bases of the banana enclaves of the Panamanian Atlàntic and Pacific have disappeared with the death of the old contracts, so bitter and painful, through which our fruit had left for so many years" (Gabrielli 1988, 13).

Indeed, splendid isolation was a thing of the past, and this remained true during the 1990s banana crisis, despite conditions that threatened to tip the balance of power back to the TNCs again. One ironic outcome of the 1990s dispute is that the UPEB, which was born out of Latin American unity in the 1974 banana war, ceased to exist after 1999, a victim of the absence of such unity in the 1990s war.

The Associate Producers during the Banana's Third Stage

The outcome of the 1974 banana war also affected the status of the associate producer programs established during the industry's second stage. Those programs were controlled by the companies, which dealt with the growers as individuals and prevented them from acting collectively. Political leaders in the exporting countries made the associate producers the focal point of their encounters with the TNCs, however, and the new parastatal institutions facilitated the formation of growers' associations. By 1976 each UPEB member had at least one such organization, which also are common in other agricultural sectors.

After 1974 the associate producer programs set out on a new path. The model for their organization was the Jamaica Banana Producers Association, which was created during the 1920s and succeeded for a while in cutting a better deal for its members. The model that emerged in the 1970s involved national and regional associations of independent banana producers who pursued the goals of eliminating intermediaries and entering directly into the transportation and marketing of bananas. They took advantage of propitious political conditions as their governments were desirous of having more of the industry's benefits accrue to nationals (Paredes 1976, 11).

One outcome of the associate producers movement was UPEB's creation of the Comercializadora Multinacional de Banano (Multinational Banana Marketing Company), or COMUNBANA, in the mid-1970s. This was a regionally organized parastatal company intended to replicate the transport and marketing functions previously handled by the TNCs. The associate producers' associations could sell their fruit to COMUNBANA as an alternative to selling it to the foreign TNCs. During the late 1970s, COMUNBANA marketed bananas from Panama to Yugoslavia (Thomson 1987, 22). The creation of COMUNBANA was an important step toward liberating the producers' associations from the control of the TNCs, but it was not a long-term success. It was unable to access enough northern markets because those markets were structured either by the TNCs or, in the case of the European Union, through longstanding relationships with former colonies. COMUNBANA went out of business in 1983, unable to survive the closing of its primary supplier, Panama's Corporación Bananera del Pacífico (Alegría 2000, personal communication).

Despite the great optimism surrounding the events of the banana war of 1974, the associate producers were unable to achieve true independence. They did pursue common goals and interests through their new organizations, and they succeeded in gaining a greater political voice than they had ever had before. But the structural aspects that have characterized the banana industry since its inception presented limits that the associate producers were unable to surmount. Clear evidence of their continued dependence emerged during the 1990s banana dispute, after the implementation of the Single European Market and a new EU banana importation policy in 1993.

Labor's Challenge during the Third Stage

The banana war of 1974 created great hopes for the industry's workers in the region. Unfortunately, the desired progress was short-lived because new political realities after 1980—both within the region and outside of it—reduced the space within which labor movements could function. Nicaragua's revolution of 1979 and ongoing civil wars in El Salvador and Guatemala changed the context for labor organizing. In addition, the Reagan administration in the United States tended to view all events in Latin America within a Cold War framework, compounding the problems faced by the region's workers. Any union suspected of Communist Party ties was repressed by forces from the political right.

The banana TNCs were quick to take advantage of the good business climate that reemerged in Central America during the 1980s. They reacted fiercely to labor problems throughout the region. One of the most confrontational situations arose in the Golfito area of Costa Rica's southern Pacific coast. UB was experiencing declining productivity there, which was later attributed to the overuse of chemicals on its plantations.[4] It refused to negotiate with the workers' union, precipitating a strike in 1984. The strike lasted forty days, during which no bananas were harvested or shipped. The government and the company both repressed the strikers, but UB's unwillingness to negotiate, because of its unstated desire to abandon the area, was what actually kept the strike going. Finally, after forty days, it declared bankruptcy, providing the pretext it needed to legally shut down its operations. Afterward, it shifted the center of its Costa Rican production back to the Caribbean side of the country. Once again the company abandoned a highly specialized economic zone to massive unemployment and poverty.[5]

In the Quepos region, also on Costa Rica's Pacific coast, UB converted many of its banana lands to plantations of African palms, which produce oil commonly used in the country's kitchens. Once planted, African palms require far less labor than bananas.

Antonio Montero, of the Instituto Centroamericano de Asesoría Laboral, noted that Standard Fruit and Chiquita conducted a *guerra santa* (holy war) against trade unions during the 1980s, stating that they equated unions with communism (Montero 1994, personal communication). The TNCs began to support *solidarismo* during that period, providing a major boost to that form of labor organization, which previously had not penetrated into the banana zones. This was the beginning of a new phase in Costa Rica's banana experience, a phase in which an ascendant solidarismo movement slowly but surely displaced the leftist unions in the banana zones, altering the unions' agendas to pursue policies and practices more favorable to management.

Many factors conspired to create the appropriate conditions for the rise of solidarismo in Costa Rica. First, the geographic shift back to the Atlantic side of the country and the subsequent expansion of the industry there released the TNCs from the system of two-year contracts that had characterized their relations with the unions on the Pacific side. Second, the move coincided with the onset of the country's foreign debt crisis. That crisis elevated the importance of export income, which was needed to help service the debt. This gave greater leverage to the TNCs, whose contribution to export income from bananas was very important. Third, the triumph of the Sandinistas in nearby Nicaragua generated fear in Costa Rica about communism generally and about the role of leftist unions specifically, reducing the country's tolerance for labor unrest. The national press figured prominently in this dynamic as it trumpeted solidarismo as an alternative to the communist-led unions. Fourth, the solidarismo movement associated itself with the Partido Unido del Socialismo Cristiano (PUSC), the less progressive of the two major political parties in Costa Rica. One of the PUSC leaders, Miguel Angel Rodriguez, wrote the law amending the national constitution to provide a legal basis for solidarismo alongside the traditional unions. Finally, solidarismo proved ideal for the implementation of the neoliberal agenda imposed on the country by the International Monetary Fund (IMF). The structural adjustment programs, a precondition for obtaining IMF bridge loans to service foreign debt, generated opposition from the country's trade unions but not from its solidarismo associations.

In 1983 banana workers voted on a referendum that offered a choice between collective (industry-wide) negotiations (a continuation of the traditional union system) and individual firm negotiations (the solidarismo option). A massive campaign followed in the country's Atlantic banana zone (Montero 1994, personal communication). The companies spent substantial sums to promote the solidarismo side and applied great pressure to influence the vote, including blacklisting union leaders and threatening their families. The unions had fewer resources at their disposal. Solidarismo won by a small margin. The defeat stripped the unions of their basic reason for existence, collective bargaining, which was not part of solidarismo. By the early 1990s, visitors to the country's banana zones could see large billboards extolling the virtues of solidarismo at the entrances to banana plantations.

The situation in Ecuador also was unfavorable to unions, despite constitutional guarantees of the right to organize. There, a national firm dominated the industry, although the North American TNCs were also present. The Exportadora Bananera Noboa (hereafter Noboa) was founded in 1952 by the Noboas, the wealthiest family in the country, who owned a shipping firm and many other companies. Noboa established plantations in the banana zones and initially sold fruit to the Standard Fruit Company. Later, in the 1950s, it entered into the transport stage of the industry and began to ship directly to the United States. Its scale of operations increased during the 1960s and again during the 1980s, when it rose to become the dominant player in Ecuador and the fourth largest banana exporter in the world, marketing its fruit under the Bonita label. It reached its peak during the banana industry's fourth stage, tapping into new markets in Western Europe, the former Eastern Bloc countries, the Middle East, and China. In 1997 it handled 13 percent of global banana exports. Its share declined to less than 10 percent after 2000, but it still ranks as the largest exporter outside of the North American TNCS (FAO 2003, sec. 6.3).

The rise of Noboa and ReyBanCorp, another Ecuadorian firm that accounts for about 4 percent of global banana exports, reflected a shift in Ecuador's banana production model during the industry's third stage, which had negative consequences for labor. The previous emphasis on small family farms was replaced by a model focusing on medium-sized farms and large corporate plantations. Today Noboa and Dole both operate plantations of their own; along with Chiquita and Del Monte, they also source more than half of their fruit from medium-sized independent growers in arrangements similar to those in Central America. The subcontracting arrangement enables them to lower their risk and absolve themselves of responsibility for larger labor forces, relying instead on the poorly paid subcontracted labor employed on the medium-scale farms. Under this arrangement Ecuador has the lowest unionization rate of all the major Latin American exporters—just 1 percent—and, not coincidentally, the lowest daily wages in the industry. Noboa actively resisted unionization throughout the 1980s and 1990s and into the new century, which resulted in the frequent strikes and violent confrontations that were documented in the European and American press (BananaLink 2006; *Houston Chronicle* 2004; *Guardian* 2002).

The End of the Empire

The Banana Empire ceased to exist by 1993. One can argue either that it met its demise in 1974 following that year's banana war or that it fell in 1958 as a result of the U.S. consent decree. Whichever timeframe one prefers, it is certain that the industry was characterized by increasing degrees of complexity as it progressed through its second and third stages. The crises triggered by Panama Disease, ongoing labor issues, the rise of new competition, the creation of associate producers programs, and the increased assertiveness of host country governments all contributed to the growing intricacy of the industry. And the complexity of the business reduced the UFCO's capacity to maintain its dominant position and preserve its empire.

The primary factor underlying the demise of the Banana Empire was the national integration that occurred within the region's nation-states, a process emanating from Latin America's stage of economic nationalism. Countries like Colombia, Costa Rica, Guatemala, Honduras, and Panama outgrew their historic core regions and expanded into their peripheral zones, which included the UFCO's banana zones. Transportation and communication were the two critical elements of this national integration process, just as they had been in the United States and Canada during the nineteenth century. As roads, railroads, airplanes, and modern communications systems penetrated the banana zones, the splendid isolation that was so carefully constructed by the UFCO slowly but surely dissipated. The actions taken by the producing countries in 1974 seemed to ensure that its reconstruction would be impossible.

TWO

The Caribbean
Banana Industries

4

Peasant Farmer Societies
Commonwealth Caribbean Bananas

The model of banana production used in the eastern Caribbean is different from the one used in Latin America. It differs in scale, landscape, land tenure systems, trading patterns, and the nature of the farmers themselves. These characteristics are a function of the relatively recent origin of the region's banana industry, which developed after World War II and during the final stages of colonialism in the region. The eastern Caribbean banana industry does share two traits with Latin America's industry, however, as both were introduced from abroad and both are dominated by foreign companies. After a brief discussion of the first Commonwealth Caribbean experience with bananas in Jamaica during the early twentieth century, this chapter focuses on the banana industries of the four small Windward Island nations of Dominica, Grenada, St. Lucia, and St. Vincent and the Grenadines (hereafter, St. Vincent).

Jamaica offers a rare example of failure on the part of the United Fruit Company. At the beginning of the twentieth century, Jamaica was the major source of the UFCO's bananas because of the Boston

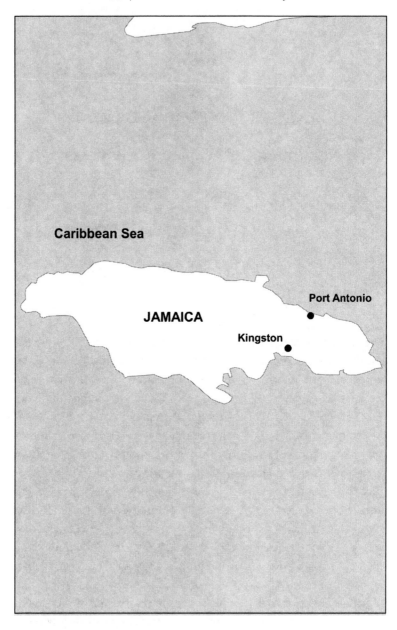

5. Jamaica

Fruit Company's earlier activities there. Later, the company's Jamaica operations were handled by Fyffes, its British subsidiary. But circumstances on the island differed from those in Central America. First, and perhaps most importantly, Jamaica was a British colony with a colonial government that believed in the rule of law and proved less corruptible than most isthmian regimes. Second, the island had a relatively good road system for that era, and because the island is small, there was no need to construct a railroad to serve the banana farms. Third, ports were open to all competing firms; no one company could gain control over the island's port facilities. Finally, Jamaica lacked remote areas that could be exploited for bananas; thus, splendid isolation was virtually impossible to achieve there. Overall, the island did not offer the kinds of conditions that had allowed the UFCO to achieve unrivaled dominance in Latin America.

Jamaica's labor situation also differed from Latin America's. A significant class of independent smallholders arose during the decades after the abolition of slavery. Most had no desire to enter into the wage labor relationship preferred by the large fruit TNCs. Instead, many joined cooperative societies that were not very amenable to the UFCO's usual monopolistic practices. In 1929, with assistance from one million dollars in government bonds, more than six thousand independent banana farmers formed the Jamaica Banana Producers Association Ltd. (Kepner and Soothill 1935, 296). Membership grew to more than fourteen thousand by 1932. The bonds were used to purchase a half-interest share in four ships, with the DiGiorgio Fruit Company holding the other half. The association purchased bananas directly from its members, paying a premium price relative to that offered by Fyffes, and DiGiorgio arranged for marketing of the fruit in both U.S. and European markets. As the activities and success of the cooperative grew, the UFCO reduced its operations in Jamaica in favor of situations where it could exercise greater control.

Jamaica's banana exports were disrupted by World War II, leading to a decline in the industry. Its postwar recovery was affected by the rise of the eastern Caribbean industries but once again involved

Fyffes, although that would subsequently change. During the 1980s, Jamaica's smaller producers were increasingly marginalized by the neoliberal policies pursued by the Seaga government (Thomson 1987, 80–81).

Although the importance of banana cultivation to the Windward Islands can hardly be overstated, the industry is relatively new. Other crops dominated earlier stages of the colonial period, when plantation monocrop systems were established throughout the Caribbean basin to serve European markets. Sugar was preeminent among the commodities that led to a nearly complete transformation of the physical and human landscapes of the region. Other important crops included cocoa, citrus, tobacco, coffee, and cotton. The topography of the four Windward Islands was not favorable for sugar plantations, and most of the estates developed there were smaller than those on the islands to their north. Labor was provided by African slaves after the demographic collapse of the indigenous population. On most islands, slaves cultivated their own food, working on Saturday afternoons and Sundays to do so, usually on small plots of land either on the margins of plantations or off the estates altogether. This contributed to a tradition of food crop cultivation in the marginal zones of several islands.

The abolition of slavery in Britain's empire during the 1830s led to great changes in the agricultural production and landholding systems of the region. Many former slaves left the estates, with some moving to towns and small cities. Others gained some form of control over the small parcels of land they had used to grow their food, often as renters, squatters, or sharecroppers if not as titled owners. These arrangements also extended to new lands, such as crown lands that were not already in use, and food was the primary output. The size of the units declined over time, through subdivision from one generation to the next.

The transition from estates to smallholdings continued into the twentieth century, stimulated by the Royal Commissions occasionally created to investigate conditions in Britain's West Indian colonies.[1] Increased competition from Cuba and elsewhere rendered most plantations in the eastern Caribbean less profitable. Many were broken up into smaller properties or redistributed to the growing peasantry, adding to the ranks of the smallholder sector.[2] Trouillot (1988, 21) offers the observation that, while peasantries predated incorporation into the modern world economy elsewhere, a Caribbean peasantry was created long after that process as an indirect consequence of it. Post-emancipation societal changes were complex and again reshaped the face of the region. They deserve greater attention than this text affords, but they warrant mention here in the context of the very different labor force and land tenure situations that existed in the eastern Caribbean at the point when the banana industry was introduced. And, since the 1950s, the region has countered the global trend toward increased size of rural production units (Brierley 1996, 1).

Declining economies and the threat of social instability caused the UK's Colonial Office and Imperial Committee on Marketing to intervene during the 1920s. The latter issued a report suggesting that a successful fruit trade could be developed between the UK and its small Windward Islands colonies, creating jobs and alleviating poor social conditions (Trouillot 1988, 124). The decision about which fruit should be selected for the new export industry was based on the suitability to a small-scale peasant model of farming and minimal start-up capital outlay requirements compared to other possible crops. The banana was chosen for these and other reasons, although implementation of the plan was delayed by World War II. It offered a relatively short period of time before beginning to bear fruit, and because it bears fruit year-round, it provided more frequent paychecks to farmers than most other commodities. The crop could be cultivated in a variety of environments, including on relatively steep slopes. In addition, a market niche existed in the UK, and Jamaica was unable to fulfill the UK's postwar demand (Thomson 1987, 4). A final factor was the ease with which the banana could blend into the diet of its cultivators. This was important because the new industry marked a shift from food production to export crop farming for peasant farmers. The fact that the export crop could also help feed the farmers was significant and shaped the nature of the industry that evolved in the eastern Caribbean. Most export crops are grown on large-scale plantations, and few fit the peasant farmer model very well. Bananas are a key exception to this norm.

Origins of the Eastern Caribbean Banana Industry

Although bananas have been cultivated in the eastern Caribbean for centuries, a banana industry has existed only since the mid-twentieth century. Two attempts to export bananas, both involving UFCO subsidiaries, occurred in the 1920s and early 1930s. The first failed due to Panama Disease and the second was a casualty of war. The current industry was founded in 1949 and took off during the 1950s as an instrument of British colonial policy. Most observers agree that the motivation for establishing the industry in the region was to quell social unrest and prevent revolutions that would threaten the privileged status of the plantocracy that still dominated island economies. There was a need to generate employment following the demise of earlier export agricultural commodities like sugar and limes. Labor strikes were common in the postwar British Caribbean, challenging regional stability within a global Cold War milieu.

Whatever the motivation, it was clear that great profits could be earned from bananas in the eastern Caribbean, and the industry there soon exhibited the trait of foreign domination that characterized the banana industry in Latin America. In the case of the Caribbean, the company involved was British rather than North American. In 1952 Geest Industries Ltd., a large British firm with Dutch origins, purchased Antilles Products Ltd., a small company that began shipping bananas from Dominica to the United States in 1949.[3] The purchase afforded Geest membership on the West India Committee, a British association that influenced colonial policy in the region (Thomson 1987, 30). Using its position on the committee, Geest convinced the

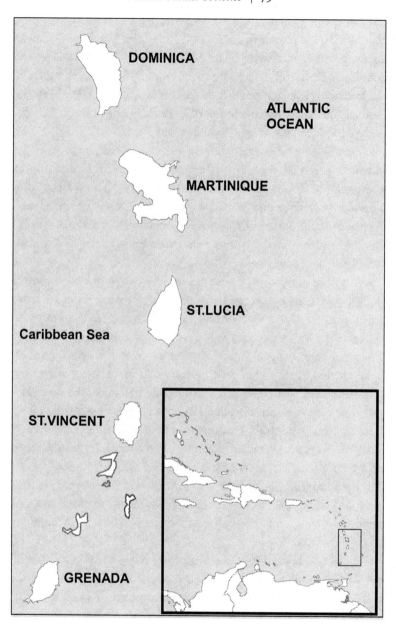

6. The Windward Islands

UK government to award it control over banana imports from the four Windward Islands in 1954. The government's willingness to do so was partially motivated by its desire to prevent Fyffes, still a UFCO subsidiary, from expanding into the region. Fyffes handled much of the Jamaica trade, and the British government preferred to keep the eastern Caribbean banana industry in British hands rather than entrust it to a U.S.-based firm (Trouillot 1988, 130).

The UK government developed a licensing system to help the industry gain a foothold in the British market. It required import companies to obtain special licenses to import bananas from "dollar-zone" producers and limited the quantities imported under such licenses.[4] At the same time, a general open license was made available to firms importing from producers in areas under British control (Grossman 1998, 39).

From 1955 until 1995, Geest enjoyed virtual monopoly control as the sole transporter and marketer of bananas shipped from Dominica, St. Lucia, St. Vincent, and Grenada. To generate the economies of scale necessary for profitability, it had to find a way to guarantee that it would supply a major share of the British market. It negotiated agreements with each island's association of banana growers for the purchase of all exportable bananas. The company was thus able to control production without growing any bananas. Instead, it used contract farming, a system involving thousands of small-scale farmers. Contract farming offered farmers the advantage of a guaranteed market for their crops and allowed capital (Geest) and the state (the UK) to maintain control over the production process (Grossman 1998, 4). Geest's contracts with the growers' associations stimulated a rapid increase in banana production throughout the Windwards.

By the late 1950s, Geest was supplying more than 60 percent of the British market. Fyffes provided most of the rest from Jamaica and Cameroon. Both operated within a protected UK market, a situation that continued after Britain joined the European Community in 1973 and signed the first Lomé Convention in 1974. During the

1960s, however, the industry recovered in Jamaica, leading to over-production, declining prices, and lower profits. This stimulated a battle between the Geest and Fyffes, with each trying to increase its market share. Resolution was achieved through a 1966 government-brokered agreement that awarded 52 percent of the market to Fyffes and Jamaica Producers, from their land in Jamaica, and 48 percent to Geest, from its Windwards sources. Both major companies agreed not to source elsewhere unless unusual circumstances, such as hurricane damage, prevented them from filling their quotas (Grossman 1998, 43).

Peasant Farmers and Social Structure

The labor unrest that began during the 1930s continued in much of the Commonwealth Caribbean into the 1970s, the decade during which the four Windward Islands gained their independence. The instability hastened the breakup of many remaining estates. Small parcels were awarded to new owners, often former workers on the original estate. This again increased the ranks of yeoman farmers, as they are called in the region. In Grenada, during the revolutionary period from 1979 until 1983, state farms replaced several of the estates. Beginning in 1986, those farms were divided into small (2.2 hectares on average) model farms, providing land to landless persons, many of them women and younger men. This system was based on a model developed in St. Lucia's Roseau Valley in 1983 (Brierley 1996, 13–18).

Numerically and production-wise, peasant farmers, rather than wage laborers, dominated the industry and agriculture in general. Although agricultural wage labor does exist, it is usually a temporary option drawn on by peasant farmers with insufficient family help at specific times during the year. Grossman (1993, 351) discovered in St. Vincent that many smallholders themselves also do wage labor to augment family incomes, with one or more family members seeking off-farm employment at least on a part-time basis. This

8. A small eastern Caribbean banana farm and farmhouse in Dominica.
9. "Dominican Gothic": Independent farmers and their families form the bedrock
of eastern Caribbean societies.

strategy is a function of small landholding size and is a continuation of, rather than a departure from, subsistence-level consumption patterns (Trouillot 1988, 268).

Most eastern Caribbean bananas are not raised as monocrops. Many farmers grow food crops as well, often in the same fields with banana plants, which provide shade for other crops. The smallest-scale growers are the most likely to use the intercropping method. In St. Vincent many farmers place bananas and food crops on the same slopes but not side-by-side (Grossman 1993, 360–61). This is likely the result of stewardship factors such as pesticide and fertilizer use. Bananas were grown in higher locations to limit the exposure of food crops to the pesticides applied to banana plants while allowing surface runoff to carry fertilizer residues downslope so that food crops would also benefit from their application. While some of the food crops are for subsistence, many are marketed locally or exported to drier islands north of the Windwards.

The life of a peasant banana farmer is not easy. Many work difficult parcels of land that would not be cultivated elsewhere. The mountainous terrain of the four Windward Islands necessitates farming on slopes. Historically, flatter land was controlled by larger units, particularly plantations, and was unavailable to the smallholders. Windward Island soils generally are productive due to their volcanic origins, but fertilizer use is common today. In addition to facing the normal challenges associated with agriculture, the farmers of the eastern Caribbean face frequent storms. The region is perpendicular to the path of the trade winds, which carry immense tropical storms, many of them hurricanes, in its direction. Hurricane David (1979) devastated Dominica to such a degree that it is still the yardstick used to evaluate contemporary storms. The 1990s were not kind to the islands, particularly Dominica. The country experienced two major hurricanes and one tropical storm within a five-week period in 1995. Most of its banana crop was wiped out and most of the island's infrastructure was damaged.

Despite the difficulty, banana farming has strong appeal to the

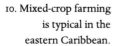

10. Mixed-crop farming is typical in the eastern Caribbean.

many thousands of participants in the industry. In St. Vincent, Grossman (1993, 355) identified factors that contribute to farmers' decision to cultivate bananas. The factors include wild price fluctuations for food in local markets, which serve as a disincentive to produce food crops. The price fluctuations are partly caused by increasing consumer preferences for imported, processed foods, a trend also noted by Thomson (1987, 62) in St. Lucia. Ironically, the imports are often cheaper than fresh foods produced locally because of the economies of scale achieved by U.S. agribusiness and because of U.S. agricultural export subsidies. These factors make it difficult for the farmers to compete. Also, state policies have favored export crops over food production to enhance foreign exchange generation. In the eastern Caribbean, such policies predate independence, extending back to British postwar policy. On the positive side, the Geest contract offered stability not available with food crops. The farmer could anticipate his or her income for a one-year period, if the crop came in as expected. Weekly or biweekly paychecks from the island's banana growers association also eliminated long periods without income, which was another plus. Most crops provide neither regular payments nor the marketing infrastructure already in place for bananas. Any new crops that farmers might plant would require another marketing system, which would be beyond the means of individual peasant farmers (Martin 1995, personal communication).

Today the number of banana farmers is decreasing in all four countries—a noticeable trend since the implementation of the Single European Market (SEM). The decline is a source of concern because it indicates falling confidence in the industry's future. It may also be a function of the high average age of the region's farmers and the fact that farming is not very appealing to young West Indians, the great majority of whom are much better educated than their parents.

An understanding of the reality faced by the banana farmers of the eastern Caribbean is critical to an appreciation of the importance of globalization and the outcome of the U.S.-EU banana war.

11. A Grenadian farmer: The average farmer in Grenada is over the age of fifty-five.

Nearly all of the banana farms in the eastern Caribbean are owned by the peasant farmers who till the land. Unlike the wage earners employed on banana farms elsewhere, these people have a direct stake in the success of the industry. The more than fifty-six thousand people working in the industry in the early 1990s formed the backbones of their national societies, generating wide income dispersion and a trickle-down effect in the industry (Sandiford 2000, 11). To a significant degree, they provide the front line against the kind of social revolutions feared by the British architects of the region's banana industry during the 1950s. Their well-being is critical to the stability of the four Windward Island societies and of the Caribbean generally. The EU noted this in the rationale for its policy, but the United States' complaint to the WTO ignored the potential regional effects of the industry's collapse.

The Banana Growers Associations

Independent farmers acting alone cannot ensure that their fruit reaches consumers in northern countries. In Latin America, where large corporations produce and handle most of the fruit, this is not a major issue. But in the eastern Caribbean, where so many smallholders are involved in the industry, a different model was necessary. This need led to the formation of banana growers' associations, commonly referred to as BGAS, in each of the four countries. In 2000, BGA membership numbers ranged from just fifteen hundred farmers in Grenada to seven thousand in St. Lucia's four associations (Reid 2000, personal communication); Dominica's and St. Vincent's BGAS each had six thousand members.[5] Operating on behalf of their members, the BGAS fulfill many functions that would be impossible for individual farmers to accomplish on their own. They also play a coordination role and serve as a collective voice for farmers at national and international levels.

The BGAS of the four countries share many traits. They are statutory agencies that were constituted by colonial governments and then reconstituted by their respective parliaments after independence. Thus, they are parastatal organizations with the public responsibility for regulating the industry while also serving members' needs. BGAS absolved individual farmers of the need to handle marketing arrangements by negotiating the shipping contracts with Geest from the 1950s until 1995, when it sold its Caribbean operations to a consortium of Fyffes and the four Windwards' governments.

In addition to shipping arrangements and generating payments to the farmers, the BGAS do bulk purchasing of the inputs for the industry, sometimes on a regional level, to get better volume prices. Inputs include fertilizers, pesticides, herbicides, hermaticides, boxes, white lime, rope, and blue vinyl covers treated with fumigants (Cyrus 1995, personal communication).[6] These items are sold on credit and applied to the *cess*, the term used for the sum deducted from the payments made by the BGAS to the farmers after their fruit has been delivered (Durand 1996, personal communication).

To cut costs to farmers, BGAS also arrange crop insurance at group rates for all four islands through the Windward Islands Crop Insurance Program (Cyrus 1995, personal communication). The insurance covers damage from high winds and hurricanes. The farmer's premium is deducted from the price received for his or her bananas. After major storms, settlement payments are based on the farmer's production average over the three years before the damage. The EU provides additional funds to help farmers recover from losses due to storms (Durand 1996, personal communication).

Extension services are provided to farmers by the BGAS, often in conjunction with the national ministry of agriculture, to help improve fruit quality. During the 1990s, the issue of quality rose in importance because of changing conditions in European markets, where nearly all eastern Caribbean bananas go. Caribbean bananas are exposed to greater competition in the EU, and consumer quality demands are passed on to the farmers.

BGAS also address their respective islands' infrastructure needs, such as the packing sheds and buying depots where farmers take their fruit after harvest. These point-of-purchase facilities are where inspection and grading of bananas occur if the bananas are not trucked directly from farm to port (Grossman 1993, 350). In addition, BGAS build a sense of community and well-being among the farmers; they hear the farmers' complaints, provide prizes or incentives for productivity and performance, and generate confidence in the future of the industry. Whereas in the past their concerns were focused on the prices Geest paid for fruit, farmers' recent complaints center on the fact that farmers are increasingly asked to do more for less. These complaints specifically relate to new packing regulations instituted in response to EU quality demands (Cyrus 1995, personal communication).

Cooperation among the BGAS was coordinated by the Windward Islands Banana Growers Association (WINBAN), an umbrella organization founded in 1958. WINBAN's major responsibility was the negotiation of the shipping and marketing contracts with Geest. It represented the region's growers in London and coordinated research efforts directed at improving the industry (Thomson 1987, 35). In 1995, after the Geest sale, WINBAN became the Windward Islands Banana Development Corporation (WIBDECO), but its basic functions did not change.

A lack of internal unity has been a problem for the BGAS, primarily because of the different farm sizes represented. The numerical dominance of small-scale farmers in the four countries does not translate into dominance in determining the direction of the associations. In fact, the owners of larger farms are more powerful, if not absolutely dominant, in the affairs of the organizations. In St. Lucia the issue of association democracy was a factor in the breakup of the St. Lucia Banana Growers Association (SLBGA) after the St. Lucia industry was privatized in 1999. The owners of many larger farms opted to break with the St. Lucia Banana Corporation, the successor to the SLBGA, and form new growers' associations (Jean-Pierre 2000, personal communication).

The Geest–Banana Farmer Relationship

Despite the efforts of WINBAN and the BGAS, the relationship between Geest and farmers in the Windwards was often problematic. The firm occupied a dominant position in the price determination process, which was difficult to challenge and detrimental to the farmers. WINBAN and the BGAS had a very restricted bargaining position because Geest could always threaten to pull out, leaving a vacuum that would be hard to fill, particularly given the company's multiple roles in bringing bananas to British tables. Geest's network within Britain would be difficult to replace. Thus, anger toward Geest was tempered by a combination of gratitude for its presence and fear that it would leave.

The true nature of the relationship between Geest and banana farmers was the source of considerable debate. Ostensibly, the affiliation was one of mutual dependency between seller and buyer of bananas. But was this really the case? Research suggests otherwise. Based on his analysis of the legal framework and contract that bound farmers to the company in Dominica, Trouillot (1988, 149) suggests that the supposed sale of bananas from the farmer to Geest did not actually occur. The island's Banana Act of 1959 obligated its banana farmers to sell all exportable fruit to the BGA. Through WINBAN, the BGA subsequently entered into a contract with Geest to which the farmers were not a party but which obligated them to sell all exportable bananas to that company alone. This dual-stage arrangement eliminated the independence of supposedly independent farmers. According to Trouillot, the farmers were reduced to proletarians who sold their labor to the company, often at less than its true value, while simultaneously bearing most of the risks related to the industry. The farmers' associations, throughWINBAN, were to have represented their interests in London and elsewhere but were ineffective in doing so. By granting Geest a "cover of legitimacy," their presence benefited the company more than farmers (Trouillot 1988, 145).

Much of this criticism is based on the pricing system in the Geest-BGA contracts. The company negotiated its banana purchasing contracts in a manner that based the prices it paid on the prices it received for the bananas in the UK. Prices were adjusted on a weekly basis and essentially worked backward. Geest, as the shipping company, deducted both the costs it incurred between Windward and UK ports and its profit margin from the price it received at the point of delivery to the ripener, often a Geest subsidiary. The remainder was the price paid to the growers' associations. After 1984 the company pricing system more closely reflected UK market conditions but was even less related to production costs in the Windwards (Thomson 1987, 31–37). The actual price paid was communicated to the BGAS after the supposed sale of the bananas occurred, effectively eliminating any bargaining that the BGAS might have done. The persistence of this system reflected the power imbalance between WINBAN (with the BGAS) and the company (Trouillot 1988, 146), at times making WINBAN look like an agency acting on behalf of the transnational corporation. The inequality was also manifested in the company's ability to reject fruit based on ill-defined quality standards. The contracts stipulated that the company only had to accept bananas of exportable quality and was not required to pay for any bananas it did not accept.[7] This is not uncommon; it also occurs in dealings between the TNCs in Latin America and the independent farmers from whom they source. But it offers one more example of the transfer of risk from the large corporation to the individual farmer.

The fact that the Windwards' banana farmers simultaneously produced food crops is a key aspect of the industry's survivability. The otherwise marginal situation of most of the region's farmers was endurable because they effectively subsidized the industry by cultivating food for subsistence, enabling the reproduction of its primary operatives (Trouillot 1988, 160). Thus, unpaid subsistence farming permitted the farmers to survive despite the low prices Gees paid for their bananas. Without this unpaid contribution, the industry would have been unsustainable and Geest's profits would have declined.

The Basic Peasant Farmer Production System

The Latin American and Caribbean banana industries were controlled by foreign interests characterized by vertical integration. In the eastern Caribbean, the dominant company, Geest, did not enter into the production stage of the industry the way the TNCs did in Latin America. Through its contracts with the BGAs in each of the Windward Islands, Geest avoided direct participation in the risky stage of banana cultivation. Those risks were borne by the region's farmers.

Geest did control the international shipping, ripening, and wholesale stages of the industry. After gaining a monopoly over the shipping of eastern Caribbean bananas in the mid-1950s, it developed a network of subsidiary firms to handle the ripening and wholesaling of bananas in the UK, earning additional profits from those operations. It sold a small percentage of its fruit to independent ripeners in Britain but handled the bulk in its own ripening facilities. The bananas spend five to eight days in these specially designed plants, where ethylene gas is used to stimulate ripening. Nearly 20 percent of the eventual price charged to consumers was earned at this stage (Thomson 1987, 57). As its final step, Geest's fleet of more than six hundred trucks distributed the bananas to retail centers throughout the UK.

The importance of organization to banana farming should be obvious. In the eastern Caribbean, networks controlled by Geest in connection with WINBAN and BGA activities provided the organizational framework within which the industry functioned. This system evolved over time, and should the industry fail, time would be required to establish a new system for any activity that might be chosen as an alternative to bananas. The entrenched system makes it increasingly difficult to wean the region away from its current primary export crop.

Beneath the organizational framework, however, banana farming in the eastern Caribbean resembles agricultural settings everywhere. It operates on a cycle, but it is conditioned by the ability of the plant to bear fruit year-round rather than just once or twice each year. Scale and terrain are important to banana production in the region. Most of the thousands of small-scale farms that dominate the industry are just 1.5 to 5 acres in size, with few larger than 50 acres (Martin 1995, personal communication). Work is done largely by hand because the topography, with its steep, narrow valleys, prevents mechanization. Hand labor lowers productivity yields, however, contributing to the higher production costs that necessitated protection in the EU market.

Because bananas do best in well drained soils, the land must be prepared for planting and good drainage systems must be developed. Plant stock is selected and the rhizomes (from which new banana plants are propagated) placed into holes dug into the soil, often on slopes. Afterward, the farmer continually battles weeds, nematodes that weaken the plant root systems, and a variety of potential diseases. Fertilizer and pesticides are used to enhance soil nutrient levels and boost yields, which tend to be lower than on the Central American plantations.

As the cultivated area increased, proximity to roads became increasingly important in determining the spatial distribution of banana farms. With packing done in the field, being closer to a road meant a shorter walk to the truck while carrying boxes filled with bananas. Road conditions are critical in determining the quality of the fruit when it arrives in port. Bananas are easily bruised; ill-maintained or unpaved roads contribute to reductions in quality ratings and, subsequently, to farmers' incomes.

The rhythm of eastern Caribbean banana farming is governed by the weekly call of the ship from the UK. The boat arrives first at Grenada and then stops at St. Vincent, St. Lucia, and Dominica. Two days before its arrival in each port, farmers begin harvesting bananas that are sufficiently mature for shipment. They are field packed and driven directly to the port by individual farmers or carried to local transport centers for pickup. A system of weighing and quality checks, managed by the BGA, is in place at both locations. This weekly ritual leads to the issuance of a paycheck, usually for fruit delivered one or two weeks previously. The frequent payments are uncommon with other crops, which makes banana farming more appealing and dissuades farmers from switching to the alternative crops that governments might promote as part of a diversification effort.

The boxes used are of a standard metric size—15 kilograms (approximately 33 pounds) each, which is smaller than the 18.14 kilogram (40 pound) boxes used in Latin America. The bananas are packed by hand; large clusters are cut into smaller sizes and gently arranged in the box to minimize bruising. Boxes are then trucked to the port for loading onto the ship. No railroads are involved, removing one of the important factors in the evolution of the Latin American industry.

New technologies, consumer preferences, and quality standards have brought changes to the process of getting produce from plant stem to the port. Significantly, in the early 1980s the industry made a transition from wet packing to field packing, driven by Geest's concerns about product quality. Previously, most farmers transported their unpacked fruit to BGA facilities for wet packing, which involved cutting banana clusters off their stems, washing them in a fungicide solution, placing them in cartons, and transporting them by truck to the port. Field packing replaced the need for those boxing facilities. Farmers began to pack their fruit in the field, and instead of the fungicide used in the earlier system, they placed crown pads between bunches of bananas to prevent the spread of mold. Small-scale farmers took their boxed fruit to the nearest inland buying center, also operated by the BGA; those operating at a larger scale—seventy-two boxes (the capacity of a wooden pallet) or more—transported their bananas directly to port. The new system reduced the handling of unboxed fruit by an estimated 75 percent, lowering damage rates and thereby also reducing fruit rejection rates. The change was therefore considered positive with regard to quality standards, but it did transfer more responsibility to the farmers. Despite the additional effort required, Thomson (1987, 72) credits field packing with increasing fruit quality and leading to higher payments to the growers.

According to Grossman (1998, 29–30), agrochemicals are used intensively in eastern Caribbean banana farming because of the quality control standards mandated by the TNCs and the state (the EU) to enhance competitiveness. One of the great ironies of the banana trade dispute is that the vast majority of agrochemicals and other inputs used in the eastern Caribbean region are purchased from the United States. Thus, if the United States succeeded with its WTO complaint, which was initiated primarily at the behest of Chiquita, a major contributor to political campaigns, the result would be reduced income for U.S. exporters of chemicals, fertilizers, tools, and other banana-related products.

Trading Caribbean Bananas

The system for the shipping of eastern Caribbean bananas reveals the structural weaknesses of the industry itself. For decades, a Geest ship sailed from the UK carrying many of the imports consumed by the residents of the four Windward Islands. The ship returned laden with their exports, mostly bananas. Despite the overwhelming importance of bananas to the national economies of three of the four Windward Island countries, each of the four would alone be incapable of generating sufficient volume to merit its own weekly call by the banana boat. The islands are dependent on one another for the survival of this—or any other—industry. This reduces their leverage in negotiating the conditions under which their banana industries operate and eliminates any flexibility they might wish to have with regard to pursuing other options. They are small players in a large industry, collectively generating just 3 percent of global banana production. Their status renders them incapable of affecting the world prices for their major export, a circumstance that reverberates through their national economies. It affects social services

12. A banana ship docked at the port near Roseau, Dominica.

13. Dropping banana cartons into a refrigerated cargo hold in Dominica.

funding, infrastructure development, and other aspects of their economic well-being.

The weekly arrival of the banana boat triggers a great amount of activity in the port area. First, manufactured products from the UK are off-loaded to be trucked to shops all over each island. While the ship is being unloaded, lines of vehicles, laden with produce for export, assemble by the entry to the port facility. Non-banana crops are loaded first; the bananas come last. Banana farmers who did not leave their fruit at one of the regional centers go through processing at the port. They first pass by the shipping company's office and drive their trucks onto the inbound scale. The weight is recorded in the company's computer system, which has a file on each farmer. At the weigh-in point, a brand might be noted, but most bananas go onto the boat with no brand name attached. They will be sold to market chains having their own brand. The trucks then pull off the scale and go for unloading; they are weighed again on an outbound scale. The second weight is recorded in the file; the difference in the two weights represents the total weight of the farmer's fruit and forms the basis of the payment he or she eventually receives.

Off-loaded cartons of fruit are initially placed in a covered area for protection from the sun and premature ripening. The cartons are arranged on specially designed pallets to be carried by forklifts to the side of the ship. There they are lifted by tall cranes and deposited into the holds of the ship. The holds are organized vertically and filled from the bottom up—first in Grenada, then in St. Vincent, and then in St. Lucia. Dominican bananas are placed in the upper holds. Each compartment is closed to allow for climate control and protection from the sun. The use of container trucks is being introduced, but to be economically feasible, containerization requires precise volumes. The forty-foot containers can be filled at locations on each island and transferred directly to the ship, but because of their greater volume, cooperation among producers is needed to use them effectively (Satney 1995, personal communication). As always, timing is of the utmost importance: no more than two days must pass between the time the bananas are cut and the time they are stored in the refrigerated holds of the ship. Beyond forty-eight hours, the risk of fungus increases dramatically, threatening the eventual marketability of the fruit. In the holds, a constant temperature is maintained to prevent premature ripening.

Vulnerability in the Eastern Caribbean

The greatest strength and the greatest weakness of the eastern Caribbean banana-farming model are essentially the same: the small-scale farmer. The thousands of family farms on which the region's bananas are cultivated form the bedrock of the democratic societies that emerged after independence was achieved during the 1970s. Bananas are virtually omnipresent; there is no distinct and separate banana zone to speak of, as there is in Central America. Bananas contribute a substantial share of export earnings while simultaneously adding to the food supply. Their importance to society is widely recognized, and there is considerable political support for the industry in terms of government policymaking within the line ministries involved with agriculture, trade, and foreign relations.

While all banana industries are vulnerable to weather, disease, and price fluctuations, the Windward Islands' model is subject to additional layers of vulnerability. It must confront changing conditions in the global marketplace that go beyond mere price variations. With the worldwide trend toward neoliberal economic policies and free trade, the eastern Caribbean countries must confront challenges over which they have little control. These challenges are embodied in the activities of the WTO and the contemporary competitiveness imperative that favors industries that are able to generate economies of scale. Such trends do not augur well for the eastern Caribbean banana industry, which until recently was able to avoid direct confrontations with global realities.

Until 1993 the survival of the eastern Caribbean industry depended on a structural framework created by the UK and the EU.

British policy was responsible for the transport and marketing systems implemented by Geest. Despite the inherent inequities in the relationship between the farmer and the TNC, the relationship facilitated the growth of the industry in the absence of alternative systems. The UK offered a protected market, alleviating the need to compete with Latin America's agribusiness banana production model. Thus, while banana farmers would never become wealthy, they were relatively assured of being able to generate a sufficient and regular income each year.

5

Belize, Suriname, and the French West Indies

On the Margins of the Caribbean

Several banana industries in the Western Hemisphere are structured differently from the Latin American and Caribbean models, although those models are the most common. Belize, Suriname, Guadeloupe, and Martinique, for example, all sell bananas to Europe, but they have different production systems from those in Latin America and the eastern Caribbean. Belize and Suriname are part of the ACP group of traditional exporters to the EU. Guadeloupe and Martinique, as overseas departments of France, are internal parts of the European Union. Although the islands are geographically Caribbean, the EU does not classify their bananas as imports. This chapter explores a few of the less common industry models.

On the Caribbean Fringe: Belize

The banana industry in Belize on Central America's northern Caribbean coast reflects the country's unique cultural mix. Belize's population of just 250,000 includes people of Latin American, Afro-

Caribbean, indigenous, and European origin. Formerly British Honduras, Belize is the only officially English-speaking country in Central America, although more than 40 percent of its people speak Spanish at home.[1] The country is a member of the Caribbean Community (CARICOM) but participates in Central American summit meetings. It increasingly functions as a bridge between the two regions but finds itself caught in the middle where bananas are involved.

Belize's banana industry has passed through distinct stages, and the country has had an interrupted experience as a banana exporter. Its initial foray into the industry began in the late 1800s, while it was still an underdeveloped colony, and lasted until 1937. During that stage, the UFCO controlled production and marketing, and Belize's industry resembled those of its Central American neighbors. At its peak, more than ten thousand acres—far more than today—were planted in bananas, mostly in the South Stann Creek district (Barry and Vernon 1995, 62). The United States was the primary market, and the UFCO exacted land and tax concessions from the colonial government using the strategies it employed in Latin America. In 1937 Panama Disease and government closure of an unprofitable railroad caused the company to shut down its Belize operations.

Several studies of Belize's banana industry were conducted over time, each leading to changes in the nature of the industry's organization. The studies began in 1968, when UB and its then-subsidiary Fyffes were invited by the government of the colony to determine the feasibility of recreating a banana industry there. The study focused on the poor southern districts of South Stann Creek and Toledo; it recommended a target size of three thousand to four thousand acres, beginning with sixteen hundred acres in the area around Cowpen. It also recommended use of irrigation systems and government support for road and pier development (Shoman 1987, 1). Fyffes declined to become directly involved in production, however, preferring to limit its investment to management of office facilities, thereby also minimizing its risks. This approach was

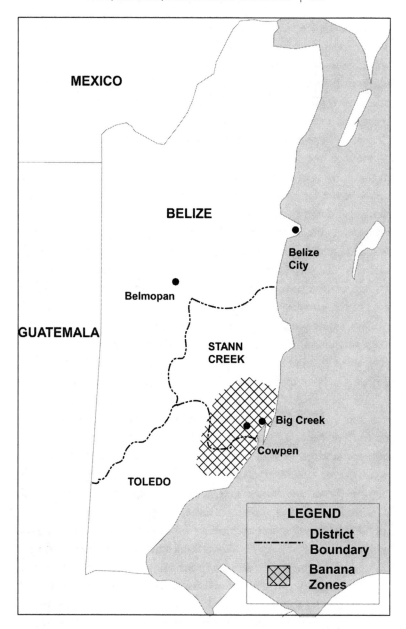

7. The banana zones of Belize

typical of UB's relatively new operating policy in LDCs at the time. The company preferred to earn its income from the services that it would provide to the industry, including technical services, shipping, marketing, and sales of inputs like packing materials, seeds, chemicals, and fertilizers. The government filled the production gap by creating a series of tenant farms, totaling sixteen hundred acres, in Cowpen. It created a parastatal agency, the Banana Control Board, to manage the industry. This state-run system was implemented in the early 1970s, and it quickly experienced two problems. First, many of the farmers selected to run the tenant farms were inexperienced and inefficient; many incurred start-up debts that they could not pay, leaving the state to take over direct operation of their farms. A second problem involved shipping. Belize's modest exports were first carried to Honduras; from there they were shipped to Germany for subsequent transfer to the UK. This circuitous route added four days to the journey, increasing spoilage losses (Peter 1982, 13).

Direct state participation in the relatively unprofitable production stage of the industry led to debt accumulation and stimulated reappraisals of the industry. The debt eventually rose to fifteen million dollars and was difficult to bear once Belize became independent in 1981 (IICA 1996, 9). A Fyffes/UB post-independence study recommended the investment of thirty million dollars to expand the industry. The funds would permit an increase in cultivated areas from sixteen hundred to four thousand acres to yield the three million boxes deemed necessary as the annual threshold for a viable industry (Benns, Webb, and Stover 1981, 3, 22). The IMF, on the other hand, criticized the level of state involvement in the economy and recommended its disengagement from unprofitable activities, including bananas (CEPPI 1995, 9). A 1984 World Bank study noted several industry limitations, including a relatively poor soil base, the cost of providing the necessary transportation infrastructure, the small number of jobs created by the industry, and the potential of alternative crops. The study questioned whether or not the industry should be continued, a question that still has legitimacy today (Shoman 1987, 4).

14. Sigatoka negra, a dreaded banana plant disease in Belize.

In response to these reports, the government privatized Belize's banana industry in 1985, moving the industry into its third stage. The decision was encouraged by UB's sale of Fyffes. The sale precluded Fyffes from competing against UB in dollar-zone countries, leading Fyffes to reorient itself toward handling ACP bananas. The government sold its banana holdings to private buyers, and Fyffes assumed management of two large farms in the Cowpen area. The Belize BGA, a private entity, has controlled the industry since 1991, when it assumed the Banana Control Board's debts and the roles the board previously fulfilled in running the industry. In return, the BGA was exempted from paying income taxes and was granted the right to charge an export tax of just thirty-five cents per box, effectively representing a government subsidy of the industry (Vix 1996, 10).

When the BGA assumed leadership of the industry, it immediately began work to control Sigatoka negra, a disease that afflicts banana plants, which was ruining the fruit and contributing to the

15. The Belize BGA's modest headquarters near the port at Big Creek.

failure of Belize's state-run industry. The intervention yielded dividends, with productivity increasing from 280 boxes per acre in 1990 to 650–700 boxes in 1998. Total production grew despite the fact that the amount of land under cultivation remained unchanged. Belize's centrally administered Sigatoka program is considered a model in the region. Soil management also improved under BGA leadership.

Belize's BGA has an exclusive contract with Fyffes for shipping its bananas, a ten-year rolling arrangement in which each year's price negotiations lead to a one-year extension of the contract. Prices paid to farmers are set during the negotiations. BGA director Zaid Flores (1998, personal communication) describes the contract as "financially strong," adding that the BGA has a good partnership with Fyffes and does not plan to change it. The BGA and Fyffes each own a 50 percent share of the Big Creek port facility from which all Belize bananas are shipped. The government limits other revenue-generating activities at Big Creek, however, to prevent competition with the main port at

Belize City. This increases the enclave-like nature of Belize's banana zone. In the southern Stann Creek district, 45 percent of all employment is directly or indirectly generated by bananas.

The Big Creek port facility opened in 1990. It replaced a smaller facility where bananas were loaded onto barges for shipment to Honduras, where they were then transferred onto oceangoing vessels. Today, in a harbor deepened by dredging, larger banana ships leased by Fyffes call once a week, usually on Monday. Fyffes issues a quota to each of the fourteen banana farms prior to the ship's arrival. Each week, cutting begins on Sunday and continues through Tuesday. Packing sheds operate at full tilt during the two-day period. The bananas undergo EU-mandated quality inspections, which include measurement for length and circumference. The fruit is checked for bruising—a serious problem due to the poor-quality dirt roads between farms and the port. Rejected fruit is boxed and sold locally or given to schools, but farmers are not paid for it. Loading by temporary workers starts when the ship arrives and continues until 10:00 p.m. each day. Because Belize's shipments are limited by the small size of its banana industry, the ship also calls at Puerto Cortes, Honduras, to obtain sufficient volume to make the twelve-day voyage economically viable (Forman 1998, personal communication).

Whereas the Windward Islands have twenty thousand banana farms, 96 percent of which are smaller than ten acres, Belize has just fourteen, none of which is smaller than twenty acres (IICA 1996, ii). The fourteen farms represent a consolidation of the twenty-six farms that operated in the early 1990s. The consolidation facilitates management of the industry and yields a vastly different labor situation. Most farms are owned by Belizeans or naturalized Belizean citizens, but Belize bananas are cultivated by wage earners. In the early 1990s, close to 90 percent of those workers were foreigners, nearly all from Guatemala, Honduras, or El Salvador. Agricultural wages are considerably lower in those countries—just $2.50–$3.00 per day, compared to a minimum wage of $9.00 in Belize, where 40 percent of banana production costs go for labor (Flores 1998, personal communication). Technically, foreign workers are required to have work permits, but many banana workers in Belize are without documents; thus, they are an illegal and easily exploitable labor force.

The country experienced several incidents of labor unrest in the banana zone during the mid-1990s. Farm owners and the BGA resisted workers' efforts to unionize the industry, occasionally in a violent manner and with government support. Their reaction to a 1995 strike led by those trying to gain recognition for Banderas Unidas, a newly organized union, attracted the attention of Belize's Human Rights Commission. Its report documented human rights violations and exposed the poor living and working conditions endured by immigrants in the banana zone. Strike leaders were deported; other participants had their work permits canceled by the government (Heusnet 1995, 3–5). The union's problems were compounded by the relative isolation of the banana zone and the fact that few workers spoke English, making it difficult for them to gain public support. According to Roches (1998, personal communication), the primary support for the workers came from Ireland, the original home of Fyffes. An Irish NGO, Banana Watch, sent a labor-organizing team to Belize to assist Banderas Unidas in its efforts.

Several farm owners responded by establishing solidarismo organizations like those found elsewhere in Central America, but this did little to diminish worker dissatisfaction (Vix 1996, 15). Others consolidated smaller farms and reduced their work force. The deportation or voluntary departure of some foreign workers and their replacement by Belize nationals reduced the participation rate of foreign workers to 65 percent by 1998. Overall, the handling of labor problems in the banana industry tarnished the country's democratic image and honorable record as an asylum state for refugees. It also embarrassed the EU, whose defense of its banana importation policy was based partly on more favorable production conditions in ACP states than in Latin America. As Banana Watch lobbied in Brussels, the EU responded by funding a housing program for banana work-

16. Banana plantation workers' housing in Bella Vista, Belize.

ers in the Cowpen area. This too generated controversy because many of the housing tracts were located on Fyffes properties, where workers felt that the company would observe their activities.[2]

Labor issues aside, Belize's banana industry made substantial progress during the 1990s, ranking as one of the country's fastest growing agricultural sectors during that decade and rising to second place (after sugar) among its export earners (cso 1995, 7–8). This success, however, yielded a new problem: overproduction (Belon 1998, personal communication). Belize was able to produce more than its EU quota of fifty-five thousand tons, but this was still not efficient enough to compete on world markets, and many in Belize see it as being locked out of those markets by the dollar-zone TNCs. The industry faces other challenges besides market access. Its productivity levels remain well below those of Latin American countries. Several African ACP competitors devalued their currencies, rendering their exports relatively cheaper on global markets. Finally, Belize must import the cartons used by its industry because it still falls short of the five million boxes per year needed to sustain a domestic carton industry.[3]

Richard Reid (1998, personal communication), senior trade economist with the Belize Ministry of Trade and Industry, was optimistic about the survival and future growth of the industry. He foresaw new opportunities in the U.S. market, facilitated by the NAFTA-induced development of Mexico's highway network, which would permit the trucking of Belize bananas directly to the United States at lower costs. This would allow Belize to ripen the fruit before shipping. He suggested that Belize could be more active in exploring options for processing bananas into derivative products, using fresh fruit rejected due to quality standards to do so. The diversification mix could include chips, baby food, dehydrated fruit, and breakfast cereals. There is wide agreement among those involved with the industry that Belize needs to produce (and market) one hundred thousand tons of bananas per year for the industry to be sustainable (Flores, Parham, and Roches, all 1998, personal communication). The production capability exists but is superfluous without the market access needed to sell additional bananas. During the course of the 1990s, Belize succeeded only in increasing its EU quota from forty thousand tons to fifty-five thousand tons per year, leaving the country far short of its desired target. Given the natural and political constraints surrounding the sustainability of Belize's banana industry, the minimal contribution the industry offers to the country's labor picture, and the damage to its international reputation caused by labor problems, the desirability of maintaining the industry can be legitimately questioned.

A State-Run Banana Industry: Suriname

How do globalization and neoliberalism affect a state-run industry? Suriname, a country of 430,000 people located on the north coast of South America was the only country in the hemisphere that still had

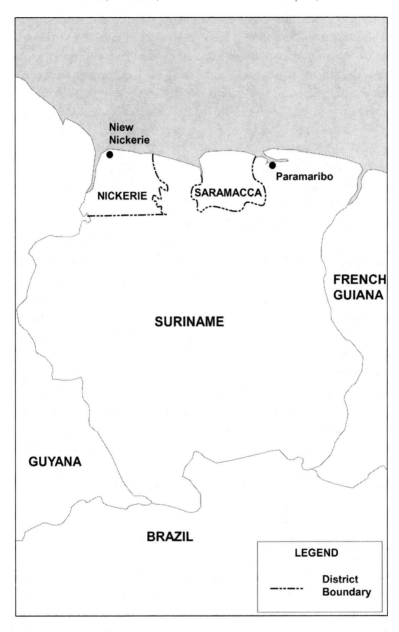

8. Suriname

a state-operated banana industry during the U.S.-EU banana war. The survival of Suriname's banana industry is critically important to its future development in the face of the country's courageous 1994 decision to reject two Southeast Asian logging firms' requests for massive forestry concessions. Nearly 90 percent of Suriname is covered by forest that is mostly inaccessible (Sizer and Rice 1995, 7). The ability to earn foreign exchange in other sectors would validate its path of rainforest conservation.

Like Belize, Suriname does not quite fit in with its geographic surroundings. It is bordered by English-speaking Guyana on the west, French Guiana on the east, and Brazil on the south. Its northern coast faces the Atlantic Ocean. Suriname was known as Dutch Guiana during the colonial period and remains officially Dutch-speaking, although most of its diverse population speaks Sranang Tongo, a Creole language. The country is culturally distinct from its Latin American neighbors and also differs from English-speaking Caribbean countries, although it is now a CARICOM member and its banana exports go to the UK, not to the Netherlands. Independent since 1975, it was politically isolated until the mid-1990s because of instability following a military coup in 1980. The restoration of democracy in the 1990s reduced the isolation, allowing the country to confront its many economic challenges. The World Bank ranks Suriname as the world's seventeenth wealthiest state, comparing its raw materials resource base (e.g., soils, forests, and bauxite) to population size. But much of this wealth remains unexploited and most Surinamese are relatively poor. A good education system, including a focus on languages, might give the country an edge as modernization proceeds (Bundel-Griffith 2001, personal communication).

Surinamers are exceptionally diverse. More than half the population is of Asian origin, descendents of indentured workers brought in by the Dutch following the abolition of slavery in 1863. The Chinese came first, but today they represent just 3 percent of the population. South Asians followed and most stayed on when their five-year contracts expired. Known as Hindustanis locally, they comprise more than one-third of the population, and Hindu temples and Islamic mosques dot the landscape. Between 1890 and 1939, thirty thousand Javanese were brought to Suriname from Indonesia, which was a Dutch colony at the time. One-sixth of Surinamers are of Javanese origins, adding to the Islamic presence. Enhancing the diverse cultural mix are predominantly urbanized Creoles (35 percent) of African or mixed African and European ancestry. In the interior, Amerindians (2 percent) and Maroons (9 percent) are the major groups, the latter being descendents of African slaves who escaped from the plantations of the coastal zone and were able to live independently in the savannah and Guiana Shield rainforest zones lying to the south of the coastal lowlands. The Maroons' independent nature was evident during the 1980s civil war, which was mostly fought in the interior. Maroons comprised the primary opposition to the military regime then in power (Bundel-Griffith 2001, personal communication). Small numbers of Europeans, mostly Dutch, and Lebanese round out this remarkably diverse population. For the most part, the various groups mix easily and the country is free of the ethnic strife that characterizes neighboring Guyana. But each group remains culturally distinct and the system of political parties is ethnically based (European Commission 1995, 20).

Suriname's physical geography hinders infrastructure development and renders the infrastructure very expensive to maintain. The country is dissected by numerous sizable rivers flowing north to the sea from the Guiana Shield. East-west transport routes were difficult to build; Paramaribo, the capital, was not linked by road to the eastern third of the country until the 1999 opening of a bridge over the Suriname River. Suriname was perhaps fortunate to have the Dutch, with their vast water management experience, as their colonial rulers. They applied their expertise to Suriname during the colonial period, draining land and developing polders on much of the fertile but low-lying coastal zone. Protective dykes were constructed to keep out the sea, allowing northern Suriname today to be a highly productive agricultural region where a variety of food and export crops, including wet rice and bananas, are cultivated. Soils are fertile but high in clay content, adding to the continual need for good drainage management systems.

In 2001 bananas were Suriname's leading agricultural export, surpassing rice. Only shrimp and fish earned more foreign exchange among the products overseen by the Ministry of Agriculture, Animal Husbandry, and Fisheries. The commercial cultivation of bananas in Suriname was preceded by research at an agricultural experiment station in the 1960s to determine whether bananas could succeed there. Despite the clay content of the soils, which is more favorable for rice, the results of the research indicated that the fruit could be cultivated successfully, and plans for the industry were launched n the late 1960s. Surland Ltd., a parastatal company, was created in 1971 to run the industry; the Ministry of Agriculture owned controlling shares in the venture, and the Ministry of Planning was a minority shareholder (IICA 1996, 10). Plantation locations in the Nickerie and Saramacca districts were selected, and from the early 1970s the industry was entirely focused on just two large state-run plantations with a total of twenty-three hundred hectares, of which just nineteen hundred hectares were cultivated annually. The industry was small by world standards but significant to Suriname (Sahtoe 2001, personal communication).

The level of infrastructure on Surland's plantations is impressive; it is similar to what one finds in Latin America. The farm at Jarikaba in Saramacca has an extensive system to bring bananas to the packing facility. Long pulley lines are activated by a motor, suspended from a virtual forest of metal A-frames, that pulls long rows of hanging bunches of bananas—as many as two hundred bunches at a time—into the packing plant. The bananas are cut into store-sized bunches and placed into washing bins. A conveyor belt removes bad fruit, which is collected for use as animal feed, so little is wasted (Amatraesijot 2001, personal communication). Export-quality bananas are washed in water and packed into plastic-lined boxes ranging in size from 12 to 18.14 kilograms.

17. Flooded fallow banana fields at the Jarikaba plantation in Suriname.

Surland uses an unusual rotation system. Individual fields are left fallow and flooded every seven years to restore the nutrient base of the soil, improve soil structure, and decrease nematode problems (IICA 1996, 25). The clay soils are not as good as silt for banana cultivation, so a moisture balance must be achieved. Clay becomes waterlogged when there is too much precipitation but bakes very hard when there is too little. Thus, effective drainage and water supply infrastructure is essential. The plantations are on polders created during the late 1950s and are laced with many kilometers of drainage ditches. The Jarikaba farm has five hundred kilometers of under-tree pipes, the largest such system in the world, which deliver water directly to the rows of plants through ninety-five thousand small sprinklers (Vlijter 2001, personal communication).

While Suriname remains dependent upon the Netherlands for markets and foreign aid, its banana industry is linked to the UK. As in Belize, Fyffes was invited to be the shipping and marketing company when the industry was initially organized. Fyffes used three or four ships at a time. Ship call frequency varied, which affected the length of the harvest period. Beginning in 2001, weekly calls prompted three-day harvest periods and more regular work for temporary labor. Shipping patterns were altered in 1998 after completion of the bridge over the Coppename River between Paramaribo and Nickerie. Previously, a small, empty ship came from the UK and called at both Nieuw Nickerie and Paramaribo. The opening of the new bridge permitted rapid road transport from the Nickerie farm to Paramaribo, so the Nieuw Nickerie stop was eliminated. After 1998, ships went from the UK to the Netherlands to take on cargo to be imported into Suriname. They called at Paramaribo, unloaded, and took on bananas before going to Costa Rica to load more bananas for the UK market (Amatraesijot 2001, personal communication). The new system increased the appeal of Suriname's bananas to Fyffes because it could derive earnings by shipping cargo in addition to bananas.

Another appealing feature of Suriname's banana industry is that, as a state corporation, Surland has goals beyond earning a profit, which it often fails to do. It fulfills an educational mission by bringing thousands of school children to its plantations each year to learn about bananas, the company, and the importance of bananas in Suriname. More importantly, Surland is a significant employer in a country with a small population. The company has sixteen hundred full-time workers and hires another eight hundred for its harvests. It employs many women, who work in the packing facilities, in the fields, and in the plant stock facility. It pays a decent wage to its workers and provides good working conditions to its unionized labor force. Working conditions appear to be far superior to those on most Latin American plantations. Facilities are clean, and pavilions are provided at each work area for people to take their breaks, have lunch, and so on. This approach yields a more costly labor force and makes it difficult for Suriname to compete on a price basis outside the protected EU market. Privatization of the industry was under

18. Surland's packing plant offered good conditions for employees in Suriname.

consideration in 2001, but many were concerned that potential buyers would not take such good care of the workers (Vlijter 2001, personal communication).

Suriname's banana industry experienced production problems and price instability during the 1990s. Uncertainty about its guaranteed EU market share also was a concern, and the country exemplified the ACP problem of high production costs. The likelihood of a free trade EU banana market increased concerns about the survivability of ACP banana industries, leading the EU to develop a project aimed at improving industry competitiveness in Suriname and other ACP states. Both the EU Parliament and the Council of Ministers approved funding, and the project began in 1999 and continued through 2004. It provided technical assistants to help run Surland's farms in an effort to improve production management. A second EU team worked on enhancing Surland's marketing capabilities and personnel management. The project also paid for the first half of the

Nickerie plantation irrigation system; funding for the second half was contingent upon whether or not project funds would be needed to restructure (i.e., privatize) the company. The EU preferred that Surland be privatized, but according to EU project advisor J. L. J. van der Ploeg, there was no buyer on the horizon (Van der Ploeg 2001, personal communication).

By 2001 the Suriname banana industry was in a crisis state. Prices had fallen to $6.50 per box while production costs remained at $8.00 per box. There was increasing sentiment that bananas might not merit continued state support if the industry could only succeed with market preferences that were no longer guaranteed. Nevertheless, considerable support for and pride in the industry remained. The high labor standards were a key to this support; several parties expressed indignation that Suriname's industry might succumb to competition from countries with less favorable labor practices. According to Jaswant Sahtoe (2001, personal communication), acting permanent secretary of the Ministry of Agriculture, Animal Husbandry, and Fisheries, "the story starts with labor—the big MNCS [multinational corporations] . . . think they can do business in Latin America in ways similar to 50 years ago." Others argue that efforts to compete with Ecuador based on labor costs would mean a "return to slavery" for Suriname (Van der Ploeg 2001, personal communication). A sense of desperation prevailed as the country prepared a diplomatic initiative in Brussels in an effort to save the industry. The fear proved well founded, as Surland ceased production later in 2001 due to its financial difficulties. With further EU assistance, production resumed in 2003, with exports realized in 2004. By 2006, however, the company still had not found a buyer for the privatization that the EU still desired.

French and West Indian: Martinique and Guadeloupe

The two French West Indian islands of Martinique and Guadeloupe are in a delicate position with regard to the banana dispute. They

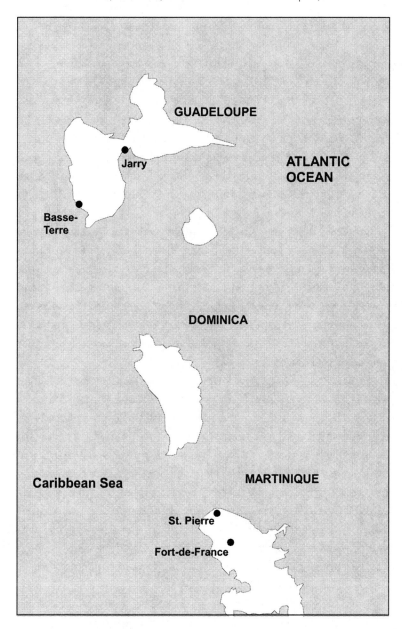

9. The French West Indies

are in the eastern Caribbean region with several ACP neighbors. However, along with French Guiana, they have been overseas departments (*departementes d'outre*, or DOMS) of France since 1946 (Burton 1995, 2), which renders them internal units of France and therefore of the EU. The EU does not classify the islands' exports to Europe as imports. French Antillean bananas are not a target of the trade dispute, despite the fact that, as a result of labor costs, they are more expensive to produce than eastern Caribbean bananas. As a result, Martinique and Guadeloupe share a concern that a truly free EU market in bananas would allow cheaper Latin American bananas to displace their uncompetitive fruit.

Banana cultivation is widely distributed on the two islands, although it is more concentrated than it is on the Commonwealth islands. In Martinique, a volcanic landmass with an irregular topography, most banana cultivation occurs in the eastern half of the island at elevations below one thousand feet. In Guadeloupe, bananas are grown primarily in southern Basse-Terre, the volcanic western half of the butterfly-shaped island, at lower elevations enabling a higher degree of mechanization.

As in the ACP Windwards, the banana industries of the French islands are not very old. Both were sugar islands during the colonial era, but in Martinique the sugar industry was not vertically integrated and it had many independent farmers. The meager profits realized from cultivation were not offset by higher profit margins in other stages of the industry (Welch 1996, 108). Along with increasing labor costs in the 1950s and 1960s, the lack of vertical integration resulted in conversion of sugar land to cultivation of bananas or other crops. A different situation prevailed in Guadeloupe, where sugar was grown by large companies that also processed the cane. Bananas gained a boost there when several of the companies planted the fruit on their unused landholdings.

Small-scale export of bananas to France began in the 1920s from Guadeloupe and in the 1930s from Martinique (UPEB 1993a, 18). Exports grew rapidly during the two decades after World War II, in response to both a rise in French demand (Welch 1996, 11) and sugar's declining terms of trade, which led to land reform in the 1960s. Less productive estates were subdivided, making additional land available for bananas, although at significant cost. This reform did not lead to independent smallholders, as it did in the English-speaking islands, because many of the landless could not afford to incur such a debt burden in a situation where very small parcels would not yield sufficient incomes to cover living costs and debt payments. Instead, it led to the rise of medium-sized farms of nine to forty-eight acres, which could yield enough earnings (Welch 1996, 56–57). Such farms became the backbone of the French Antillean banana industry, and bananas are generally not grown on farms smaller than twelve acres there.

Banana farming in the French West Indies is virtually always practiced as an intensive monoculture rather than as part of a mix of crops. The number of farms is significantly smaller and the average farm is larger than typical banana farms in Dominica, St. Lucia, St. Vincent, or Grenada. Most farmers employ non-family help, usually on a part-time or seasonal basis. Welch (1996, 35) estimates that when banana production peaked in the 1960s, there were eight thousand banana workers in Martinique and six thousand in Guadeloupe. The use of wage labor augments the importance of bananas to the economies of both islands, although neither Guadeloupe nor Martinique depends on bananas as much as the ACP Windwards do. With the advent of the SEM in 1992, Martinique had 20,500 acres under cultivation on 1275 farms and Guadeloupe had 1238 producers farming 15,600 acres (UPEB 1993a, 18).

Labor is costly in the French islands because employment must conform to the norms of metropolitan France, which include social payments. The norms mandate that workers be used as intensively and effectively as possible to justify their cost. This policy encourages a greater reliance on fertilizers to boost crop yields to levels sufficient to recover labor costs. Good wages also attract workers to agriculture, although young people are generally reluctant to work in that sector.

The French Antillean banana industry has depended on protection from France since its inception. Its initial niche in the French market was created in the 1920s and 1930s by the use of tariffs and quotas on fruit from outside France's colonial domain, a practice continued during the postwar period. Decolonization during the 1960s led to a reorganization of the French banana market, with two-thirds allocated to Martinique and Guadeloupe and one-third to the former colonies of Cameroon, Ivory Coast, and Madagascar, which produced bananas more cheaply and could have displaced the Antillean bananas in the French market had there been no quotas. Nearly 54 percent of the West Indian quota was allocated to Martinique and more than 46 percent to Guadeloupe (Welch 1996, 26–29), and these shares were preserved in 1993 when the SEM banana market was organized.

France's protection of an uncompetitive industry in two of its DOMS perhaps makes more sense when viewed in the context of the choices available. France must purchase more costly bananas, but in doing so it boosts the economies of two islands whose economic distress would become a burden of a different kind. Should their banana industries collapse, more French public funds would be required to offset the resulting recession. The loser, should one be identified, is the French consumer and taxpayer. The choice to protect Martinique and Guadeloupe can be rationalized as a cost of France's long-term policy of assimilation of its former Caribbean colonies, a policy with roots in the 1848 emancipation of slaves in the French empire. Assimilation was intended to preempt support for independence movements in the colonies. With departmentalization in 1946, the level of commitment to improving living standards on the islands increased, enhancing the importance of nonharmful economic policies in the process. As noted by French West Indian Aime Cesaire in 1947, "Indigenous separatist movements are all too often the result of a failure by the metropolis to keep its promises" (Hintjens 1995, 20).

Today the farmers of Martinique and Guadeloupe are EU citizens who are entitled to many benefits unavailable to farmers in neighboring countries. These include Common Agricultural Program (CAP) subsidies paid to farmers to offset the higher production costs that would otherwise render their bananas uncompetitive. Such payments were opposed in the 1960s by other EU states that questioned the "European-ness" of West Indian fruit.

Other funds are dispensed directly to the two islands for infrastructure improvements under EU regional development programs. The funds have led to a fine road system, including feeder roads that provide access into areas of banana cultivation. Thus, unlike Belize, the farmers of the French West Indies need not be concerned about fruit being bruised by rough trips to port along bumpy, dirt roads. Overall, such benefits amount to a per capita transfer of funds to the DOMs nearly thirty times greater than EU aid to ACP citizens in neighboring islands. This largesse contributed to stronger support for the 1992 Maastricht Treaty, which changed the European Community to the European Union, on the two islands than in continental France. While France narrowly approved ratification, with 51 percent in favor, Martinique and Guadeloupe cast 70 percent and 67 percent of their votes respectively in favor of the treaty (Jos 1995, 87). Should a free EU banana market be created, it is clear that the two DOMs face less dire consequences than their Commonwealth neighbors. Both islands enjoy free emigration to continental France, offering an escape valve not available to independent countries of the region. In addition, as departments of a larger nation, they are not forced to achieve a positive trade balance by themselves. Nevertheless, the loss of such an important productive sector in what is a heavily service-oriented economy would have a negative economic impact and yield a significant displacement of people.

On the Caribbean Margins

The effects of globalization on the banana industries in the Americas are determined by a combination of factors, including the scale and economic relationships that characterize each industry. Whether one is considering large-scale plantations in Latin America and Belize or family farms in eastern Caribbean, however, decisions made in Geneva, Brussels, and Washington have a profound impact, particularly with regard to the events of the U.S.-EU banana war during the 1990s. The institutional frameworks within which the contemporary globalization of bananas has occurred, including the European Union, the WTO, and the Lomé system that links the EU to the ACP states, are integral to the battles of the banana war.

THREE

The Changing Framework
of the International
Banana Trade

6

The Single European Market and the Western Hemisphere's Banana Industries

Bananas have been a problematic issue for the EU since its inception as the European Economic Community (EEC). The 1957 Treaty of Rome that established the organization specified that the original six member states of the European Community (EC) had until 1970 to adjust their banana importation policies to conform to the standard CAP imposition of a 20 percent tariff on foreign bananas. That outcome was not easily achieved. The Benelux states had to raise their tariffs, meaning higher prices for consumers. Italy was forced to lower the 36 percent tariff it assessed on all bananas except those from Somalia. West Germany, the EC's highest per capita consumer of bananas, was granted a duty-free banana import quota, enabling it to continue tariff-free import of the cheaper Latin American bananas favored by German consumers. Among the six original members, three—France, Italy, and the Netherlands—had banana-producing colonies that would soon become independent countries.

The organization grew from six countries to twelve by the time negotiations on the Single European Market began in the late 1980s, and the complexity of the banana issue grew with it. Britain's 1973 entry into the EC was especially significant because the UK had special trade relations with several banana-producing former colonies. The three newest members—Greece (1982), Spain (1986), and Portugal (1986)—had their own banana-growing regions. Thus, six of twelve EC members produced bananas or had former colonies that did, or both. A seventh—Ireland—was the home base of Fyffes, a large importer of bananas into Europe.

By the early 1990s the twelve members of the EC collectively comprised the world's leading banana market, increasing the importance of the industry. It was clear that developing a joint policy for importing bananas—as mandated under the SEM—was not going to be an easy task. This chapter discusses the arduous process of bringing bananas within the SEM, the policy devised to do so, reactions to that policy, and the EU's early efforts to implement the policy, which provoked varied responses on four continents.

Pre-1992 Banana Trading Policies Involving EU Members

The pre-1992 EU banana market is best described as fragmented, with each member state having its own banana trade policy. The fragmentation reflected the three major source areas from which European consumers obtained bananas and the relationship of each member state to those areas. The source areas were (and are) the EU itself, ACP countries, and "third countries" (the EU's term for countries other than EU and ACP exporters), mostly in Latin America. The twelve members' policies can be grouped into four categories. Spain stood alone in its own category, protecting its market for its Canary Islands producers. It employed a strict quota on bananas from elsewhere, importing only when demand could not be satisfied internally. France, the UK, Greece, Italy, and Portugal comprised the second category and used a system of preferential access for internal or ACP production. France protected its market for bananas from its internal producers, Guadeloupe and Martinique, and from Cameroon,

Ivory Coast, and Madagascar, three ACP exporters with whom it had a special relationship. Britain and Italy had no internal production but provided preferential access to ACP exports from Belize, Dominica, Grenada, Jamaica, St. Lucia, St. Vincent, Suriname (to Britain), and Somalia (to Italy). Greece protected its market for internal production from Crete, and Portugal did so for bananas from the Algarve, the Azores, and Madeira. The five countries imported third-country bananas to fill their remaining demand, applying the CAP 20 percent tariff to those imports (European Commission 1991).

The third category included Belgium, Denmark, Ireland, Luxembourg, and the Netherlands, all of whom imported bananas mostly from third countries and applied the 20 percent CAP tariff. Although Suriname was a Dutch colony until 1974, its banana industry was linked to the British market, not to the Dutch, and the Netherlands did not offer preferences to Suriname's fruit. Finally, Germany's policy stood alone. The original tariff exemption it negotiated as part of the Treaty of Rome continued until the implementation of the SEM. Thus, German consumers enjoyed a free market and low prices for bananas, meaning that they had the most to lose with any EU importation regime that fell short of free-market conditions. This factored into the German stance during the SEM negotiations.

Summarizing the pre-1992 situation, EU members had twelve distinct banana importation policies, six operating with protected markets of various kinds and six allowing market forces to shape the flows of banana imports. This was not compatible with the free-market ideal toward which the EU was moving. But the bananas consumed by EU citizens were produced under vastly differing conditions and at varied costs. A free EU market would favor the most cheaply produced bananas, in this case third-country bananas, of which 98–99 percent came from Latin America. It would permit Latin American bananas to displace EU production and ACP imports in a short period of time, with dire consequences for the countries involved.

Anticipating the SEM in Banana-Exporting Regions

The years preceding the implementation of the SEM were filled with some uncertainty but with more optimism in banana-exporting Latin American countries. The countries began to position themselves to take advantage of what they hoped would be an opportunity to increase their exports to the EU. This positioning involved increasing their banana production and exports to EU member states wherever possible to expand their export history, in case that history was used as a reference baseline to determine the level of post-1992 imports. The expansion occurred against a backdrop of events in Eastern Europe that also promised to yield new markets for Latin American bananas, although this did not materialize. The ACP banana exporters faced similar uncertainty and, where possible, responded in a similar manner. However, their aspirations did not extend to Eastern Europe because they had little hope of marketing their uncompetitive fruit in that region.

Table 2 illustrates banana consumption in the European Community by source category and country for the years during which the negotiations over the framework of the SEM occurred. Most ACP and Latin American states increased their exports during this period; the trend was particularly evident in Cameroon, Ivory Coast, Colombia, Costa Rica, the Dominican Republic, Ecuador, Jamaica, and Nicaragua. In those countries the 1991 figures represented sizable gains in the three-year average, raising it above pre-1989 levels. For the nine Latin American countries in the table, the 1991 figure is 15 percent higher than the 1989–91 average, indicating a surge in exports immediately prior to 1992, the EU's target year for implementing the SEM. Belize also took steps to increase its production during this period, although its output had not yet reflected the increase. Only Somalia, with its political and economic crises brewing, showed a different trend.

The increased banana production generated controversy in Latin America for several reasons. It was fueled by an expansion of the cultivated area, often at the expense of other crops, without guar-

TABLE 2. European Community average banana consumption (in metric tons)

Source	1989–91 average	1991
Total	3,316,559	3,694,619
European Community sources	637,826	636,643
Canary Islands	344,132	339,450
Guadeloupe	94,047	116,124
Martinique	199,648	181,069
ACP producers	604,608	657,735
Belize	23,412	19,616
Cameroon	83,447	115,841
Cape Verde	2,820	3,011
Dominica	52,897	54,961
Grenada	6,548	8,168
Ivory Coast	83,447	116,425
Jamaica	57,505	70,116
St. Lucia	115,387	102,651
St. Vincent	70,732	63,063
Somalia	41,783	8,177
Suriname	28,491	27,744
Third countries	2,061,031	2,394,199
Colombia	441,787	512,347
Costa Rica	522,340	569,373
Dominican Republic	4,709	10,298
Ecuador	408,975	600,766
Guatemala	28,128	13,186
Honduras	136,907	138,396
Nicaragua	47,285	65,218
Panama	470,886	484,574
Venezuela	14	41

Source: CORBANA 1993a, Cuadro no. 2

antees of access to restricted portions of the EU market. In Ecuador the territory cultivated in bananas doubled during the three-year period, leading to overproduction and a decline in prices (Van Sant 1993). The growth occurred with public-sector support in several countries and was criticized for its contributions to environmental degradation (Asociación Ecologista Costarricense 1992; IUCN 1992; and Soto 1992). The criticism increased after the EU announced its banana importation regime in 1993, which indicated the failed wisdom of the expansion programs.

Debating the New Banana Importation Policy

The negotiations over what to do about bananas proved to be one of the major stumbling blocks faced by the EU in its SEM planning efforts. The negotiation process exhibited substantial discord, reflecting the member countries' established trade patterns and priorities. The European Commission's Management Committee for Bananas oversaw the negotiations; it was mostly composed of representatives from Ministries of Trade or Agriculture in each member state (Appleton 1996, personal communication). The final plan had to be approved by the Council of Ministers, the EU's primary policymaking body that had approved the framework of the Treaty of Maastricht.

France and the UK, supported by Greece, Portugal, and Spain, led efforts to shape a policy that preserved ACP preferences and protected EU banana production. Germany spearheaded the opposing side, favoring a free market in bananas. It did not wish to see its advantageous duty-free situation altered in any way. Germany's position was supported by several Latin American countries and by the United States, whose transnationals were active in the Latin American industry.

Several proposals were debated and the many vested interests in the banana industry lobbied in Brussels and other EU capitals. In 1990 the UK presented a proposal that included a quota for third-country bananas, to be periodically increased; a licensing system to import those bananas; a tariff on such imports; and transitional subsidies to EU and ACP producers negatively affected by such a system. This proposal was opposed by the Caribbean Banana Exporters Association (CBEA), which argued that it would lead to a decline in banana prices that would be harmful to its members (Welch 1996, 324).

In 1991 France proposed a quota system for third-country imports and price controls for bananas marketed in the EU. It also proposed allocating licenses for 30 percent of the third-country bananas imported into the EU to firms that traditionally handled ACP bananas, allowing those companies to sell the licenses for five dollars per box, potentially representing a $180 million subsidy to those firms. France was supported by Caribbean interests but strongly opposed by Germany, U.S. transnationals, and Latin American producers. Germany continued to push for a free-trade system with no quotas. As a compromise, it suggested that the CAP tariff on third-country bananas be increased to 30 percent, potentially increasing EU revenues by $500 million per year (Brinard 1996, personal communication).

An accord was reached in late December 1992, shortly before the SEM was to go into affect. Among the tens of thousands of goods imported into the EU, the banana was the last product for which an agreement was struck. The approved proposal was a victory for Britain, France, and Spain and a defeat for Germany and the free-market advocates. The accord's late approval delayed its implementation until July 1, 1993, to allow the parties involved to prepare themselves for its complex provisions.

Council Regulation 404/93

Council Regulation (EEC) 404/93 provided the framework for a single market in bananas, ending more than a year of speculation about what form the new banana importation regime would take. It addressed each of the three primary sources of EU bananas, laying down the structures that would govern how each could continue

TABLE 3. Annual European Community production eligible for subsidies

Source region	Metric tons of bananas
Canary Islands (Spain)	420,000
Crete and Lakonia (Greece)	15,000
Guadeloupe (France)	150,000
Madeira, Azores, and Algarve (Portugal)	50,000
Martinique (France)	219,000
Total	854,000

Source: European Council 1993, article 12

to market its fruit within the European Community. A separate section authorized a licensing system for selected third-country imports. The policy was further refined by subsequent regulations that provided the operational details of the plan in a process that has continued since 1993. The preamble of Council Regulation 404/93 presents the EU's goals in creating the new policy. One stated goal was the survival of the EU's banana industries despite the structural shortcomings that limited their competitiveness. The new regime continued the protection and CAP subsidies previously offered to EU producers, but it limited eligibility for subsidies to an annual maximum of 854,000 metric tons (European Council 1993, article 12). At first glance this appeared to limit the size of the EU banana industries, but the cap actually surpassed the 1989–91 production levels by 34 percent, thereby protecting the industry and allowing it future growth (table 2). The breakdown of the 854,000 metric ton subsidy cap by region appears in table 3.

Traditional ACP banana exports were defined in Article 15, which established an 857,700 metric ton annual limit for duty-free access to the EU market. This total was subdivided into individual country allocations for the twelve states protected by the Lomé Convention Banana Protocol (table 4). Exports above that figure or from ACP states not among those twelve were considered nontraditional and

TABLE 4. Allocation of traditional ACP banana exports to the EU (in metric tons)

ACP country	Allocation of traditional banana exports
Belize	40,000
Cameroon	155,000
Cape Verde	4,800
Dominica	71,000
Grenada	14,000
Ivory Coast	155,000
Jamaica	105,000
Madagascar	5,900
St. Lucia	127,000
St. Vincent	82,000
Somalia	60,000
Suriname	38,000
Total	857,700

Source: European Council 1993, annex

were subjected to a tariff. The article provided no mechanism for the transfer of unused allocations among the twelve to cover situations such as hurricanes, in which production by one or more members might be disrupted. As the data in table 2 illustrate, 857,700 metric tons surpassed the export level of the twelve countries during the years preceding the implementation of the SEM, leaving room for growth. Such growth, though, would be more attainable in countries like Ivory Coast, Cameroon, Belize, and Suriname than in the four Windward Islands countries that were experiencing difficulties with their banana industries.

Finally, third-country exports were addressed in Article 18, which created an annual quota of two million tons for bananas imported from third countries or nontraditional ACP states. Third-country bananas were subject to a 100 ECU tariff per ton, while nontraditional ACP bananas included within the quota could enter duty-free.[1] The article provided for additional imports, should unmet demand remain, but a tariff would be assessed to any bananas imported above the quota—850 ECUs per metric ton if the bananas were from third countries and 750 ECUs per ton if they were from nontraditional ACP sources. This combination of quotas and tariffs is known as a tariff-rate quota (TRQ), and it quickly became one of the new policy's most objectionable features. The high tariffs were set to preserve space for more European Community and traditional ACP production in the EU market (European Council 1993, preamble). Despite the use of the third-country terminology, the tariff-rate quota applied primarily to Latin America, where 1991 exports to the EU (table 2) exceeded the tariff-rate quota by 16 percent and 1992 exports reached a record 2,398,677 tons, 16.7 percent above the quota (CORBANA 1993b). The imposition of the tariff-rate quota thus thwarted Latin American goals of increasing banana exports to the EU, offsetting the effort in the region to increase production prior to 1992. Much of the controversy surrounding Council Regulation 404/93 centered on this aspect of the policy, especially considering the inability of European Community and ACP traditional producers to fill their market shares.

The Licensing System for Importing Third-Country Bananas

In addition to delineating market shares for the suppliers of the EU banana market, Council Regulation 404/93 established a licensing system to govern the two million metric ton tariff-rate quota for third countries (Article 17). Article 19 added to the complexity of the situation by authorizing three categories of licenses (table 5). The operators that were eligible for licenses in each category were determined by their historic banana-related activity, except for the 3.5 percent of companies in category C, which was created for new firms entering the industry. The license system generated great controversy. The eligibility requirements for category A and category B licenses represented a significant departure from the pre-1992 situation and directly challenged the established power structure of the industry. The new system allocated category A licenses, which

TABLE 5. Types of import license holders

Category	%	Tons (millions)	Type of operator
Category A	66.5	1.33	Those who historically marketed third-country and nontraditional ACP bananas in the EU
Category B	30.0	0.6	Those who historically marketed European Community and traditional ACP bananas in the EU
Category C	3.5	0.07	Those who began marketing third-country and nontraditional ACP bananas in 1992 or after

Source: European Council 1993, article 19

carried the right to handle nearly two-thirds of the two million ton tariff-rate quota, to firms with a history of importing third-country and nontraditional ACP bananas into the EU, most of them subsidiaries of the dollar-zone TNCs that had handled nearly all of the more than two million tons of third-country bananas imported into the EU in the years prior to 1992. Because the vast majority of profits in the industry are generated during its post-harvest stages, limitation of category A licenses to 66.5 percent of the shipping and handling activity meant a significant decrease in the earning potential of the companies involved. Category B licenses carried the right to handle 30 percent of the tariff-rate quota bananas. These were granted to companies that historically imported traditional ACP bananas into EU member states, mostly EU-based TNCs such as Fyffes and Geest (Zúñiga 1994, personal communication). Category B firms effectively gained the importation rights to six hundred thousand metric tons of dollar-zone fruit that previously had been supplied by their U.S.-based competitors. This aspect of the new policy was a major target of the United States' complaint to the WTO.

A three-year historical reference period was used to determine who received category A and B licenses and the quantities they were permitted to handle. Commission Regulation 1442/93 (1993) subsequently stipulated that licenses were to be divided among three categories of firms: primary importers, ripeners, and free circulation releasers, who often also import or ripen (European Commission 1993b, article 3). Retailers and wholesalers who did not participate in any of the three defined activities were not eligible for licenses. A company wishing to obtain a license had to register to import bananas into one EU member state. That state's "competent authorities" determined the quantity that the company would be licensed to handle, although the company was free to sell its bananas anywhere in the EU. A company could choose the countries it wished to import from, but this ultimately had to be coordinated at the EU level, where reduction coefficients were applied to determine how much each company could import from each country (Appleton 1996, personal communication).

EU banana trade specialist Zelie Appleton (1996) described the licensing system as transparent, noting that licenses had value but no price. They were free; the EU did not charge for them. The competent authorities in each member country issued the licenses according to the regulations laid out in the policy. Once a company had a license, it could sell it, but selling a category A license decreased the quantity of a firm's future allocations because it diminished the firm's three-year trade record. Appleton considered it unlikely that much trading of category A licenses occurred.

Selling a category B license would not affect the subsequent quantity a firm was eligible to handle because the firm's ability to obtain a category B license was based on its ACP activity, not its dollar-zone trade. Thus, the sale of category B licenses was thought to be brisk, although no solid information on sales of licenses was available. Once a license was issued, the EU did not track what was done with it. U.S.-based TNCs or their subsidiaries could buy category B licenses from their original holders; no effort was made to control the prices that were charged in such exchanges. Appleton estimated that depressed banana prices in late 1995 caused the resale value of the licenses to drop to two to three dollars per box from an estimated five to six dollars per box when banana prices were high. At any price, however, the sale of category B licenses to import third-country bananas by firms that historically handled traditional ACP bananas constituted a subsidy to those firms, which might reflect the true intention of Article 19 of the policy. The European Community Banana Trade Association's Philippe Brinard suggested that the transferability of category B licenses was necessary to keep such firms active in handling the less profitable ACP bananas (Brinard 1996, personal communication).

Rationalizing the Policy

Why would the EU develop such a complicated regime to govern the importation of bananas, especially when it knew that doing so would generate a storm of controversy and resistance on several fronts? Why not create a free market in bananas that reflected the spirit of the Maastricht Treaty and the SEM? The answers to these questions are found in the complex and delicate balance of interests that the Europeans were attempting to satisfy, interests that reflected the differences among the three primary categories of producers from which the EU sourced its bananas.

Council Regulation 404/93 was an effort to address several priorities simultaneously. First and foremost, internal priorities were considered. Protection of EU banana production was part of a larger effort to raise the living standards of European Community farmers and maintain rural landscapes in one of the world's most densely populated and industrialized regions. The EU invested substantial resources to accomplish these goals through its CAP and regional development programs. This aspect of the policy was not challenged in the 1990s banana war, although EU banana-producing regions felt threatened by the United States' WTO complaint.

The new policy's role in addressing another internal priority was questioned, however. EU documents frequently refer to the goal of protecting its consumers, but it is not at all clear how Council Regulation 404/93 contributed to achieving this goal. Theoretically, a free market in bananas would lead to increased competition and lower-cost fruit. The new trade regime did not permit this in an unfettered manner. While banana prices did decline in member states that previously had protected markets, they rose in Germany and other free-market members. Where prices did decline, it is difficult to ascertain whether the decline resulted from the policy or from the market glut brought about by overproduction, which was caused by the increased production during the pre-1992 period.

ACP issues usually received second priority in EU rationales of the regime. The organization's traditional links with the ACP and Lomé Convention obligations were often noted. For example, the policy's preamble reiterated the European Community's desire "to ensure satisfactory marketing of bananas . . . originating in the ACP States within the framework of the Lomé Convention Agreements, while maintaining traditional trade patterns as far as possible" (European Council 1993, preamble). The lack of competitiveness among the twelve traditional ACP banana exporters was an additional consideration. Since SEM rules permit the free movement of goods across EU internal boundaries, cheap third-country bananas imported into a non-tariff member could be sold anywhere in the EU, bringing competition to ACP bananas in previously protected markets like the UK or France. Thus, it was argued that some guaranteed market access was essential to prevent the complete displacement of ACP bananas by cheaper third-country bananas (Matthews 1993, personal communication). Also, EU members like France and the UK spent great sums on development aid policies to assist ACP states; they had no desire to enact policies that offset such efforts.

Third on the EU's list of priorities was maintaining good relations with other regions, including Latin America. The EU operates the world's largest foreign aid program, which is directed in part toward individual Latin American countries. Nevertheless, the announcement of the new trade regime was perceived in those countries as a full assault on Latin America by the EU.

Although it might seem that the new policy would induce a shift in the geography of banana production and trade away from Latin America and toward the ACP producers, in fact the EU trade regime, if left alone, would have preserved banana geography. Had the EU created a completely free market in bananas, it would have led to lower production in EU and ACP states in favor of Latin America, with negative results for the ACP group. By offering incentives for firms to continue to operate in ACP states, rather than abandon them, Council Regulation 404/93 was actually a maintenance regime, not an intervention designed to produce a shift in banana production. The licensing system and the ability to sell category B licenses were the keys to ensuring that outcome.

Reactions to the New Policy

Predictably, a policy with such potentially far-reaching ramifications generated a wide variety of reactions, most of which can be categorized along geographic lines. The policy received support from European Community producers, EU member states with strong ties to the ACP exporting countries, importers eligible for category B licenses, and the ACP group in general (Reuters 1993c). The most vocal opposition came from Germany and Latin America. Initially, the United States was surprisingly silent on the issue. Germany, supported by Belgium, Denmark, and the Netherlands, fought the implementation of the policy by filing a challenge before the European Court of Justice, the EU institution that deals with disputes between member states and between members and the European Community itself (Reuters 1993a, 1993e). The complaint was denied in June 1993, immediately before the policy's implementation date (*European Report* 1993; Reuters 1993b).

The governments of Latin American exporters opposed the new policy. They viewed the tariff-rate quota as a serious limitation to their ability to export bananas to EU markets, especially given their inflated expectations of what the SEM would mean for them. Gloomy predictions abounded, including cutbacks in cultivation, substantial job losses, and attendant social problems. UPEB led the way, representing all of the major Latin American exporters except Ecuador. It estimated that Latin American banana exports to the EU would decline by five hundred thousand tons in 1993 and that financial losses would total one billion dollars during the following three years. It also suggested that five hundred thousand Latin American workers and family members would be directly affected through loss of employment. Furthermore, it predicted a price decline due to oversupply that would ultimately lead to even greater losses for all producers (UPEB 1993b, 7).

Rumors and a lack of understanding of the policy guided reactions in Latin America, raising questions and concerns while generating predictions of doom. In Ecuador, the world's leading banana exporter since 1952, the fruit represented 25 percent of total exports, second to petroleum (40 percent) and its derivative products (Government of Ecuador 1993, 2). Uncertainty centered on the activities of the North American TNCs, which, although they did not own land in Ecuador, purchased 43 percent of the bananas exported from Ecuador in 1991. Because Council Regulation 404/93 offered no individual country quotas for third-country producers, growers questioned whether the TNCs would continue to handle as much Ecuadorian fruit, given the greater distance from Ecuador to Europe and the fees required for use of the Panama Canal. Its regional competitors have outlets on the Caribbean, alleviating that additional cost. Ecuador felt that it would absorb much of the six hundred thousand ton reduction suffered by category A importers, including reductions suffered by several of its own national firms (e.g., Bananera Noboa; BAGNO, S.A.; and Rey Banano del Pacífico), which exported 47 percent of the country's bananas (Government of Ecuador 1993, 8). The national firms are subject to the same additional transport costs as the foreign TNCs shipping from Ecuador.

Ecuador had 210,000 people directly employed in bananas in 1991—18 percent of all rural sector employment. Bananas generated

27 percent of agricultural gross domestic product and supported related industries, including the manufacture of cartons and plastic bags and financial and transportation services (Government of Ecuador 1993, 3). A government report predicted a yearly loss of 385,000 metric tons in banana exports to the EU, half of its 1991 total and valued at $88.5 million. Over a ten-year period, the loss would equal 10 percent of Ecuador's foreign debt. The government also expected an 11 percent decrease in banana prices in 1994 as a result of oversupply, leading to an additional loss of $65 million from its exports to other markets, for a total loss of $153.5 million per year. The financial loss would prompt reductions of 126,000 acres in land under cultivation, mostly on small and medium-sized farms that would be unable to weather the expected price declines. The reductions would negate the $500 million in investment used to develop the land. The report predicted that the EU policy would be a catastrophe for Ecuador, leading to "illicit activities like delinquency and narco-trafficking" (Government of Ecuador 1993, 10).

Costa Rica was no more optimistic than Ecuador about the effects of Council Regulation 404/93. Although the numbers were smaller, expectations of cultivation cutbacks and job losses prevailed, leading to a similar array of socioeconomic ills. Predictions of banana export declines ranged from 10,838,869 boxes to 16,351,869 boxes per year, with job losses expected to fall between 4,522 and 6,522 posts, calculated at one job per cultivated hectare (CORBANA 1993a, 22–24).

The new EU banana licensing system understandably generated considerable confusion. It fueled rumors, later shown to be untrue, that the EU intended to raise an estimated $600 million per year in license fees (Sánchez 1993, personal communication). There was also confusion as to whether or not the granting of licenses to import companies constituted individual country quotas for third-country exporters. This was desirable from Costa Rica's perspective, but it was untrue. Licenses conveyed quotas to importing companies, but they did not allocate those quantities to individual countries. Rather, they reflected the traditional organization of the banana industry, in which individual companies controlled most aspects of production, transportation, and marketing in a highly vertically integrated industry (Matthews 1993, personal communication). The lack of individual third-country quotas offered no security to the exporting states and proved divisive within the region.

Costa Rica took the lead in seeking consensus among Latin American exporters to establish a unified front in combating the new policy. It moved quickly, filing a complaint with the General Agreement on Tariffs and Trade (GATT) (Reuters 1993d). Its complaint, GATT's first case involving the Lomé Convention, charged that the new policy discriminated against Latin American exporters and that it was a barrier to the free trade espoused by the GATT's principles. Colombia, Guatemala, Nicaragua, and Venezuela supported Costa Rica in its GATT case, but Ecuador could not as it was not a GATT member. The complaint caused consternation within the European Commission, which feared a negative decision and damaged relations with Latin America. The concern increased when the issue was taken up by UPEB, which included all major Latin American banana exporters except Ecuador. Ecuador's tradition of following an independent path where bananas were concerned can be explained by its lower dependence on foreign TNCs and its lower production costs, which it felt could allow it to gain a competitive advantage over its regional rivals in the EU market. Ecuador's actions frustrated Costa Rica's effort to achieve unity in Latin America.

Among ACP exporters, the reaction was more positive, although even they did not view the policy as a panacea to the ongoing problems of their banana industries. In Dominica, Grenada, Belize, and St. Lucia, the policy was welcomed as a means of buying the time needed to get their industries in order. Prices were a major concern because the new regime eliminated the favorable price structures found in the protected markets that imported ACP bananas. Dominica BGA Director Conrad Cyrus commented that the policy brought lower prices and led to a hand-to-mouth operation for most Dominican growers (Cyrus 1995, personal communication). This was because it subjected ACP bananas to European market forces from which they were previously shielded.

The reaction in Belize was mixed. The country was generally supportive of the new regime but disappointed in its relatively small quota. Its initial allocation of forty thousand metric tons per year was based on low pre-1992 production levels. According to Zaid Flores, Belize BGA Secretary General, use of that reference period did not provide sufficient time for the great strides made following the reorganization of the industry to take effect. Those gains were realized after 1992 (Flores 1998, personal communication). A 1993 study of the country's banana industry indicated that it needed to be able to market ninety thousand to one hundred thousand tons of bananas per year to be competitive. That report prompted Belize to request a higher allocation from the EU.

Implementing Council Regulation 404/93

In the midst of controversy, the EU began implementing Council Regulation 404/93 on July 1, 1993; its midyear start necessitated an interim regulation to cover the transitional period. Commission Regulation (EEC) 1443/93 accomplished this task, establishing the rules to cover the remaining six months of 1993 (European Commission 1993c, article 1).

The impacts and shortcomings of the new regime were felt quickly. Surprisingly, the ACP group experienced the greatest initial difficulties. One of the problems was the traditional ACP country quota system, which provided no mechanism for transferring quotas among the twelve countries to cover situations where natural disasters or other events prevented a country from filling its quota. By default, this left the matter to the Commission's Banana Management Committee. When Hurricane Debbie struck St. Lucia in September 1994, it destroyed much of the country's banana crop. German members of the committee blocked the transfer of St. Lucia's unused quota to other eastern Caribbean producers (*Financial Times*

1994, 1). The problem occurred again in 1995, when Hurricanes Luís and Marilyn struck Dominica and produced similar results.

The solution selected to cover ACP shortfalls due to tropical storm damage was to permit extra dollar-zone imports. An additional 150,000 metric tons were added to the 1994 tariff-rate quota, and the 1995 tariff-rate quota was raised by 100,000 tons. The licenses to handle those amounts were added to category B and were granted to the companies that suffered losses in their imports from ACP sources. Those firms were also entitled to EU indemnity payments for a nine-month period, the time required to bring in a newly planted banana crop (Brinard 1996, personal communication). In the meantime, Belize, Cameroon, and Ivory Coast wanted to make up the shortfall but were not permitted to do so (Appleton 1996, personal communication). Thus companies seemed to matter more than countries in this decision-making process.

Price convergence within the EU was another problem encountered by ACP producers after 1993. As predicted, banana prices rose in countries like Germany, Belgium, and the Netherlands, but they fell in the UK and France, the countries that purchased the bulk of the traditional ACP bananas. This drove many small-scale farms in ACP states to the brink of infeasibility because the farmers received lower payments for their bananas. The problem contributed to social unrest in the eastern Caribbean, where the economic impact of banana farming is greatest. The BGAs in Dominica, St. Lucia, and St. Vincent provided subsidies to their members, expenditures that those associations could ill afford.

In Belize the BGA's effort to increase its allocation was partly successful in 1994, when its annual quota was raised to fifty-five thousand tons. This quantity still fell short of the amount needed to sustain a viable industry, leading to problems of overproduction, the dumping of excess bananas, and a restructuring of the industry. The latter began in 1996 and involved making "the difficult decisions" (Flores 1998, personal communication). Zaid Flores explained that the BGA was left with two choices. It could carry on with twenty-six growers and continue being inefficient, or it could reduce the number of farms. It chose the latter option and began closing down the least productive acreage and reducing the number of farms from twenty-six to fourteen. The smaller farms, which had the highest break-even points, were the ones most affected. Overall, twelve hundred out of fifty-two hundred acres were taken out of production. The fifty-five thousand metric ton quota, equivalent to three million forty-pound boxes, was distributed among the remaining farms.

Within Latin America, the EU policy led to worsening relations among the major exporters, particularly between Colombia and Costa Rica on one hand and Ecuador and Guatemala on the other, destroying any hope of presenting a united front on the issue. But even more damaging was the tendency of individual countries to respond to the new regulations by competing with one another for a greater share of the reduced EU market for third-country bananas. Their efforts were oriented toward reducing the costs of their bananas and forging alliances with companies holding the coveted import licenses. Their actions undermined the gains realized by Latin American states during the third stage of the industry's evolution, including the increased tax revenues from the industry.

Costa Rica's response was typical. It took measures to deal with what it foresaw as an imminent crisis. It sought to diversify its markets for bananas, looking at Russia, Eastern Europe, and Asia, although those did not prove to be lucrative outlets during the 1990s. In August 1993 it lowered the banana export tax from fifty cents to thirty cents per box, its lowest level since 1975 (Decree No. 22335-H-MAG, 1993). This was done to maintain the country's attractiveness to the U.S. TNCs and EU importing firms. Unfortunately, the decrease applied to all banana exports, rendering the decline in public-sector revenues more significant. Projected losses from the lower tax were $25 million for 1994, more than half of the sum collected in 1992 (CORBANA 1993b). Additional reductions in public-sector receipts were expected from the tax on imported inputs used in the Costa Rican industry. Lower production and export volumes meant reductions in the quantity of imported inputs and an anticipated further loss of $1.6 million in government revenues, based on an average of ten cents per box of bananas exported (Zúñiga 1994, personal communication). These actions threatened to return the country to an earlier era when it derived few public-sector benefits from one of its major export commodities.

In summary, the introduction of the new EU policy clearly illustrated how policies made in one region affected countries and people in other parts of the world. In the long run, many of the early dire predictions did not materialize, but the anticipation of severe consequences was sufficient to propel the banana dispute into the world's primary international trade forum at the time: the GATT. This occurred virtually at the same time as the transition within the global trade milieu that led to the formation of the GATT's successor, the WTO, which would then spearhead the globalization process in the banana industry and other economic sectors.

7

Neocolonialism Encounters the Free Trade Imperative

The Lomé Convention has figured prominently in the international banana trade. Named for the capital of Togo, where it was signed in 1975, Lomé at first linked nine EC members to forty-six of their former colonies in Africa, the Caribbean, and the Pacific. The latter comprise the ACP group, now numbering seventy-nine states, including many of the world's least developed countries but not including Latin America's mainland countries. The Latin American countries were not included in the ACP group because they have stronger economic linkages to North America than with Europe.

Since 1975 there have been four Lomé Conventions; the first three were five-year agreements, and the fourth, Lomé IV, lasted ten years before expiring in February 2000. The future of the system was in doubt during the last five years of Lomé IV as negotiations for a fifth convention were held simultaneously with WTO hearings on the United States' complaint against the EU banana policy. Those hearings clearly affected the outcome of the Lomé deliberations, which yielded a successor, the Cotonou Agreement, signed in 2000.

Trade preferences were among the major provisions of the Lomé Conventions, although each convention was multifaceted and also included provisions not related to trade. As colonies, the ACP states exported nearly all of their primary sector commodities to EU members; the accord preserved this pattern by guaranteeing preferential access to EU markets for those and selected other goods. Because the system provided an avenue of continued European influence in its former colonies, it was often criticized as neocolonial by structuralists and world systems theorists. This criticism is justified, but the fact remains that the ACP countries relied heavily on Lomé's provisions for economic stability and middle class employment. It was the ACP group that provided the impetus for renewals, with adjustments, of the convention. In the 1990s the GATT, which had granted waivers for Lomé's trade preferences, was replaced by the WTO, which at first also granted waivers but made it clear that these were to be used only temporarily, while a new system was developed. The WTO viewed Lomé as a barrier to free trade.

The Origins and Philosophy of Lomé

By 1995 the Lomé Convention was the world's largest system of North-South cooperation, having passed through an evolutionary process that reflected the vastly changing geopolitical, economic, and social realities in the countries involved. Its roots, however, extend back to the beginning of the European Community. The 1957 Treaty of Rome that established the EEC also created the European Development Fund (EDF) to provide development assistance to less-developed parts of the world.[1] The rapid decolonization that occurred after 1957, especially in Africa, led the EC states to seek new methods of association with their former colonies. In 1963 they and several African states signed the first of two Yaoundé Accords in Cameroon's capital. The primary intent of the Yaoundé Accords was to assist with infrastructure development in the newly independent states, mostly former French colonies. Some trade benefits were also conferred. Today the Yaoundé agreements are credited with establishing the precedent of European cooperation with the ACP group of countries.

The 1973 EC accession of the UK, with its associated Commonwealth of Nations, stimulated a change in the scope and nature of Europe's relationship with former colonial territories. The UK had an established tradition of trade preferences for its colonies and ex-colonies and wanted this to continue within the EC framework. By the mid-1970s, several Caribbean and Pacific colonies gained their independence and needed to be considered in future agreements. These changes were embodied in the first Lomé Convention (1975), which formally created the ACP group and initiated what is now called ACP-EU Cooperation. Nonreciprocity is a key aspect of the relationship. Theoretically, the benefits flow south, particularly to the least developed countries in the ACP group, whose limited industrial exports could enter EU markets duty-free.

Adaptability was a key element of the Lomé system. Each convention covered a finite period, allowing adjustments and revised aid packages to be incorporated. The revisions permitted the system to respond to vast changes that occurred after 1975 in the greater global political and economic frameworks. The Lomé Conventions sought to enhance the development levels of the ACP states. They addressed development in a multifaceted manner that included programs, projects, and mechanisms that targeted most sectors of each nation's economy. The system grew increasingly attentive to human development, reflecting the greater global attention paid to issues such as gender, food security, population, and the environment. Since its inception, Lomé was, both in theory and practice, an interstate system that functioned as an intergovernmental organization (IGO). Like other IGOs, including the United Nations and the African Union, it operated through an intergovernmental consultation process to develop its policies (European Commission 2000a, 4). Assistance passed from one government to another; civil society was not directly involved. The intergovernmental process was later viewed as a shortcoming of Lomé, and civil society began to emerge more forcefully to fill the gaps left by unresponsive governments in many ACP states.

The first Lomé Convention covered a five-year period and was funded at approximately $3.6 billion. It focused on industrialization and initiated the nonreciprocal preferential access for many Europe-bound exports from ACP countries—a Lomé trademark. As is common in many agreements today, Lomé I made reference to the equal relationships among the parties involved, their mutual interdependence and respect, and the rights of each country to develop its own policies. In 1975, however, respect for former colonies was a rather recent phenomenon that reflected optimism about their future.

The second Lomé Convention, signed in 1979, covered the period 1980–85. It offered $5.4 billion to fifty-eight ACP states. The agreement addressed issues related to mining industries that were economically important to several ACP states and provided funds intended to reduce dependence on the mining sector. Lomé II negotiations also included discussions about investment guarantees and human rights; although neither ultimately figured into that accord, the discussions set the stage for serious subsequent consideration of those issues (European Commission 1996, 2).

By the time Lomé III was signed in 1984, the ACP group had expanded to include sixty-five countries. The third convention covered the 1985–90 period, a time of famine in several African countries. It was apparent that development needs were still great and that basic human needs often remained unmet. As a result, priorities under Lomé III shifted to food security and attainment of self-sufficiency in ACP countries. EDF assistance grew to $8.9 billion. The convention also emphasized the service sector for the first time, providing support for tourism and maritime transportation.

The fourth Lomé Convention covered a ten-year period beginning in 1990. Its funding, however, was for five years, with the stipulation that the agreement would undergo a review at that time. The midterm review allowed for needed adjustments and a new financial protocol to fund programs during Lomé IV's second half. When the fourth convention began, the ACP group had sixty-eight members and the EC was preparing to become a union with the 1993 implementation of the SEM. Lomé IV emphasized neoliberal structural adjustment as applied by the World Bank and the IMF during the late 1980s to many indebted countries of the South (European Commission 1996, 2). Its policies involved economic restructuring, opening up restricted economies, removing trade barriers, downsizing state sectors, and privatizing many state-run entities. Such policies were criticized globally for their impact on the world's poor; Lomé IV sought to ameliorate their worst aspects by including social dimensions in its program.

The midterm review of Lomé IV had to account for the vast changes that occurred during the first half of the 1990s. These included the disintegration of the USSR, the end of the Cold War, structural adjustment in Eastern Europe, instability in the Middle East, democratization in many ACP countries, the EU's implementation of the SEM, penetration by the GATT into agricultural trade, and the transition from the GATT to the WTO. All of these events and processes influenced EU policymaking in ways that affected the ACP-EU relationship. The ACP again increased its memberships, reaching seventy members by 1995.

The midterm review introduced a greater degree of conditionality into Lomé's finances by linking funding to respect for human rights, the rule of law, and democratic principles (European Union 2000, 2). The funding changes reflected changing international realities, increased EU commitments to other regions, and growing impatience on the part of EU taxpayers, who wished to see greater results from development aid. The revised accord promoted intraregional ACP cooperation, environmental protection, economic diversification, improved status for women, and a greater role for the private sector. Lomé IV was funded at levels of $14.4 billion (1990–95) and $17.5 billion (1995–2000). Taking inflation into account, this was the first time that cooperation funding did not grow in real terms.

Lome's Mechanisms

The Lomé Conventions articulated several diverse, ambitious goals for achieving socioeconomic progress in some of the world's poorest countries. Fulfilling those goals required new procedures and mechanisms, some of them designed specifically for ACP-EU cooperation while others were variations of earlier themes. They included an administrative structure, commodity protocols, trade preferences, and the STABEX system.

The Lomé system was institutionalized but not heavily bureaucratized, although it had to deal with bureaucracies in the countries that were party to the conventions. It operated through programmed dialogues that drew participants from institutions in participating countries and IGOs. Participants included EU and ACP legislators, the ACP-EU Council of Ministers, and the Ambassadors Committee. The primary permanent institution, the ACP Secretariat, was based in Brussels, the headquarters of the European Commission, and provided support to the participant dialogues.

Today ACP states maintain diplomatic missions in Brussels. Smaller states that are unable to afford their own missions establish joint missions in Brussels. CARICOM represents several small Caribbean states there. The EU maintains delegations in ACP states that function much like embassies, and they are led by a chief of delegation responsible for implementing cooperation programs (called National Indicative Programs) within the Lomé framework. These EDF-financed programs were negotiated by the European Commission and each ACP state for five-year periods corresponding to the EDF budget programming cycle. They incorporated economic and social program targets and identified the EU financial commitments needed to achieve those goals.

Trade preferences were among the cornerstones of the Lomé system and a source of controversy during the 1990s. Preferences removed impediments to the flow of ACP goods into the EU but did not apply to goods flowing in the other direction. Despite global trends toward free trade and the application of equal trade policies to all partners, the revised version of Lomé IV (1995) did preserve several preferences. These included exempting selected ACP goods from tariffs applied to the same products imported from third countries (Article 168) and absenting ACP goods from quotas, except for products covered by special protocols (Article 169). In addition, the accord prevented the EU from adopting trade policies harmful to ACP states without prior consultation. The EU used this provision to defend its banana importation policy before the WTO, citing its inviolable commitment to the ACP group as the policy's rationale.

The four conventions included several commodity protocols as annexes to the general agreement. Each focused on a specific commodity (e.g., bananas, sugar, or beef) and defined special rules for the importation of that commodity into the EU. Protocols applied to specific countries and did not generically include all ACP states. They also established quotas from each country covered by the protocol, defining the quantity of the commodity eligible for special treatment. That special treatment included price supports (sugar), import tax refunds (beef), and duty-free access to the EU market (bananas). The Caribbean region primarily benefited from sugar, banana, rum, and rice protocols, of which the sugar agreement was the most significant in terms of value (CARICOM 1996, 45–46).

The Lomé accords also instituted the STABEX system in the 1970s. This unique approach recognizes one of the worst legacies of the colonial experience for many ACP states: unpredictable price fluctuations among many primary sector commodities. The restructuring of colonial economies to meet the needs of metropolitan powers left most LDCs dependent upon a narrow range of agricultural and mineral exports. Independence did not change this reality and it exacerbated the problem of unreliable prices, compromising foreign exchange earnings in many countries and making it difficult to implement development programs from one year to the next. Lomé sought to overcome this problem for agriculture by establishing the STABEX fund. The fund is made available each year, but its disbursement is not guaranteed. It is intended only to cover those instances where funds are needed to offset the negative impacts of foreign exchange shortfalls caused by price declines for a country's major commodity. Payments, referred to as "transfers" in EU documents, directly target the affected sectors or substitute goods for a diversification effort (ACP Secretariat 1995, article 186).

Through STABEX, an ACP government could be reasonably certain that stable income levels would be maintained, allowing it to plan social and economic investment expenditures. Although the system was intended to overcome price fluctuations, its use was extended to cover losses in export earnings caused by natural disasters. That proved critical during the 1990s, when Caribbean storms destroyed bananas and other crops in several ACP states. A similar concept was applied to ACP mining sectors through the Lomé System for Minerals (SYSMIN). SYSMIN loans were first used during the 1970s, when a special financing facility was created to implement the system (ACP Secretariat 1995, 56).

During Lomé's twenty-five years, many EC funds available to ACP states went unused, not because they were not needed but because of the cumbersome procedures used to request funds, the restricted nature of EDF funding, and the failure of ACP institutions to claim the resources to which they were entitled (European Commission 2000a, 5). By 2000 approximately 9.9 billion euros remained uncommitted (European Commission 2000b, 3). In the negotiations for Lomé's replacement, this was one area identified as in need of improvement.

The Banana Protocol

The Banana Protocol was the fifth protocol in the revised Lomé IV, forming one of the bases of the EU's defense of the banana policy. Article 1 of the protocol states: "In respect of its banana exports to the Community markets, no ACP state shall be placed, as regards access to its traditional markets and its advantages on those markets, in a less favorable situation than in the past or at present." The EU interpreted this statement to mean that it could do nothing to harm the banana industries of those ACP states defined as traditional exporters of bananas to EU member countries. The great lengths to which the EU went to preserve Council Regulation 404/93 indicates how seriously it viewed this commitment.

By 1995, however, it seemed clear that the preferences conferred by the Banana Protocol would become a casualty of globalization and free trade. The major thrust of the Lomé IV revisions of 1995 was to identify and implement measures needed to enhance the competitiveness of the traditional ACP banana export industries. It made specific reference in Article 2 to aspects of the industry needing attention, including production quality, research, harvest, post-harvest handling, packaging, storage, transportation, marketing, and trade promotion. The list left out very little, indicating that the industries in question still had a long way to go to achieve competitiveness. Article 3 mandated formation of a permanent oversight committee to monitor progress in attaining the required improvements. Although the protocol did not refer to specific funds, the traditional exporters were eligible for STABEX and other funds, and the EU invested considerable sums to help the ACP twelve make their industries more competitive. The twelve traditional ACP banana exporters covered by the protocol were entitled to the preferences in Council Regulation 404/93. Their classification as traditional distinguished them from other ACP banana exporters, which did not receive the same benefits. The EU classified those exporters as nontraditional because they had begun only recently to export bananas to the EU. The two most significant nontraditional exporters were Ghana and the Dominican Republic. Ghana was a charter ACP member but did not historically export bananas to the EU. The Dominican Republic's situation was more complex. It signed the third Lomé Convention in 1984 and was at the time the only Latin American ACP country. The twelve traditional banana exporters agreed to its accession only under the condition that it be classified as a nontraditional exporter—an indication of the value of the traditional classification to those twelve countries (Appleton 1996, personal communication).

The ACP banana exporters, particularly those from the eastern Caribbean, viewed the U.S. challenge to the EU policy as a direct assault on the Banana Protocol. Sir Neville Nicholls, president of the Caribbean Development Bank, voiced concerns in 1994, when the Office of the U.S. Trade Representative (USTR) launched an investigation of EU banana trade preferences. The lines of the banana battle were drawn when Nicholls identified Chiquita and the Hawaiian Banana Association as the motivation behind that quest (*New Chronicle* 1994a, 2). His statement was indicative of commonly held sentiments that the subsequent U.S. challenge in the WTO was actually written by Chiquita lawyers.

The Demise of Lomé Foretold: Limbo in the Late 1990s

By the late 1990s the future of the Lomé Convention was in doubt. The end of the Cold War and the steady march of globalization changed the milieu within which the system operated. The ongoing banana dispute threatened the continuation of preferential trade arrangements that were a hallmark of Lomé. Furthermore, in 1996, during the preparation stage for new negotiations, the European Commission issued a green paper that analyzed the Lomé system and identified several shortcomings that would need to be addressed in any new agreement. Based on more than twenty years of experience (at that point) with trade preferences, the green paper concluded that trade preferences had not induced economic take-off in developing countries and that Lomé had failed to increase trade among ACP states, including within the regional ACP blocks. The analysis also showed that imports from the ACP group had declined since 1975, from 6.7 percent to 3 percent of total EU imports. Overall, just four ACP countries—Ivory Coast, Jamaica, Mauritius, and Zimbabwe—evidenced successful macroeconomic growth indicators (European Union 2000, 6–7). Also, by the late 1990s, many European taxpayers questioned the wisdom of providing assistance to undemocratic, corrupt governments, particularly in Africa. They doubted "the viability and effectiveness of cooperation" when coop-eration produced disappointing results in situations where "insuffi-cient account had been taken of the institutional and policy context in the partner country" (European Commission 2000b, 2).

These factors did little to generate optimism about Lomé's future, particularly among ACP states, as the expiration of Lomé IV loomed on the horizon. For many, February 2000 signaled the end of an era—an era of preferences and protection. As early as 1995, doubts were expressed about whether a new Lomé Convention could be ne-gotiated. All anticipated a major retreat from the principles of Lomé, if any agreement at all were to be reached. In Dominica, Grenada, Belize, and St. Lucia, pessimism reigned, particularly with regard to continued trade preferences. In Dominica, the foreign ministry's Sonia Magloire-Apgar indicated that CARICOM was handling the issue of Lomé's preservation for the smaller Caribbean ACP states, but she saw the results as uncertain (Magloire-Apgar 1995, personal commu-nication). The banana case was always mentioned as an indicator that free trade was taking precedence over the idea of fair trade em-bodied in the Lomé arrangement. Gregoire and Fadelle stated that Dominica's National Development Corporation was operating on the assumption that Lomé would *not* be renewed and was working to ensure that the country would be able to replace the 50 percent of its foreign exchange earnings it expected to lose from the demise of its banana industry following the expiration of Lomé IV (Gregoire and Fadelle 1996, personal communication).

Harold Parham, of the Belize Ministry of Agriculture, described as "hopeless" the efforts to preserve all of Lomé, but he hoped that some aspects, such as STABEX, would remain. He said that Belize wanted to work with the EU to create new strategies to assist the ACP states in their efforts to survive in the globalizing economy (Parham 1998, personal communication).

Despite the general pessimism, export sectors within many ACP countries appeared to view February 2000 as an effective dead-line. This was certainly true among banana producers in the eastern Caribbean and Belize, where the pending expiration of Lomé IV stimulated efforts to achieve competitiveness. According to the Dominica Banana Marketing Corporation's Tim Durand, becoming competitive meant improving quality and yields to the point where producers could consistently deliver a high-quality product cost-effectively (Durand 1996, personal communication). He saw this as a difficult goal but added that they must try to attain it. He also noted that the small size and sweet flavor of Dominica's bananas should enable them to carve out a niche market, alleviating the necessity of competing on price alone.

Amidst this climate of change and pessimism, in September 1998 the ACP group began to negotiate with the EU to prepare a new agreement. The process was difficult and often strained, but the negotiators persevered. Their deliberations continued into February 2000, concluding a few days before Lomé IV's termination date. Many factors contributed to the challenge they faced, but it was clear that the banana trade issue and the WTO response to it affected the negotiations and that any new agreement would have to be WTO compliant.

The Cotonou Agreement: A New System

The new accord that emerged was called the Cotonou Agreement, named for yet another African capital—in this case, the capital of Benin, where it was signed in June 2000. Replacing Lomé rather than extending it, Cotonou was presented as "a new partnership" among its ACP and EU signatories, and it embarked on an unprecedented twenty-year term, with reviews every five years (European Commission 2000a, 3). The agreement signed in Benin reflected the changing milieu within which the negotiations over the future of the Lomé system occurred. On the surface, its rhetoric sounded like the EU-ACP relationship was proceeding at full steam, with an agenda that, rather than signifying a retreat, was overly ambitious in attempting to combine progressive ideals with neoliberal policies.

A careful reading of its provisions, however, reveals that Cotonou recognized Lomé's shortcomings and the realities within which the new system would operate.

Cotonou addressed many of the concerns about Lomé, including human rights and refugee issues, democratization, sustainable development, poverty eradication, promotion of private investment, and better integration of ACP states into the world economy (ACP Secretariat 2000, 5). The new system also promises greater involvement by civil society and NGOS. Article 4 states that "the parties recognise the complementary role of and potential for contributions by non-State actors to the development process." Thus, it is clear that the parties to Cotonou sought to satisfy a different set of conditions from those governing the Lomé treaties.

Aspects of the new system that reflect the globalization and neoliberalism shaping today's world are evident even in the Cotonou articles dealing with social and human development issues. While the articles are consistent with Lomé efforts to overcome the negative effects of colonial rule and the harmful impacts of structural adjustment, the lack of specificity in Articles 25–27 may be an indication of the EU's reticence to directly confront the neoliberal policies responsible for those impacts.

Cotonou's development sections also reflect changing times and a confusing set of priorities. There is frequent reference to sustainable development, which by the 1990s had risen to near-paradigm status in development planning, and there is support for rural development strategies that establish a "framework for participatory decentralized planning" and micro-enterprise development (ACP Secretariat 2000, 22). Article 56 advocates the promotion of "local ownership at all levels of the development process," while elsewhere the promotion of foreign investment is prioritized. Other themes reflected in various Cotonou articles include the promotion of public-private sector dialogue and cooperation, the development of entrepreneurial skills and business culture, privatization, and enterprise reform.

Article 22 mandates support for structural adjustment and reaffirms the right of ACP states to "determine the direction and the sequencing of their development strategies and priorities." The two emphases seem contradictory because structural adjustment is usually imposed by foreign concerns, including the IMF. Article 23 includes a promise of cooperation for the promotion of fair trade, despite the fact that Cotonou overall seeks to prepare ACP states to compete in a free-trade environment. Despite a sizable EU financial commitment, expected to total more than eighteen billion dollars, Cotonou exhibits conflicting priorities that could yield disappointing results.

Cotonou does not include specific funds for the STABEX and SYSMIN systems, but it does preserve the concept embodied in those Lomé mechanisms in Article 68. Instead of permanent STABEX or SYSMIN funds, support for Cotonou will be determined on an ad hoc basis and funds will be drawn from other EDF sources.

The WTO looms as a noteworthy but ominous presence in the accord, particularly in those sections pertinent to trading arrangements. Cotonou does not capitulate outright to the WTO, but it recognizes the organization's importance and does not challenge it. Rather than an outright surrender, it represents an effort to buy time to allow the ACP group to better prepare for the free-trade rules of the future. This is what the EU has been trying to accomplish all along with bananas. Cotonou extends that approach to other sectors as well.

But the need to be attentive to WTO demands is clear. Article 34 presents the objective that economic and trade cooperation must "be implemented in full conformity with the provisions of the WTO," while Article 35 stresses the importance of "enhancing ACP States' competitiveness," a WTO priority and an obvious necessity if free trade prevails. Article 36 deals with the modalities of Cotonou's trading arrangements. It commits the signatory parties to concluding WTO-compatible trading arrangements, but it establishes a preparatory period to do so (scheduled to end no later than December

31, 2007). Toward that goal, it retains the nonreciprocity embodied in Lomé, but only during the preparatory stage. It also "reaffirms the importance of the commodity protocols" while agreeing on the need to review the protocols within the framework of wto rules. Article 39 encourages greater participation by ACP states in the wto. Many still are not members of that organization. Rules for the post-preparatory trading arrangements, determined through ongoing EU-ACP negotiations, would be implemented no later than January 1, 2008. A review was scheduled in 2006 to determine whether countries would be ready for the new rules or if more time would be needed. The high level of specificity with regard to dates was necessary as a means of gaining wto approval for a waiver for Cotonou, allowing the EU to extend many aspects of the Lomé system for several years without having to constantly defend them in court.[2]

With regard to commodity protocols, Article 40 states that the "parties recognize the need to ensure a better operation of international commodity markets and to increase market transparency." It is clear that the Cotonou negotiators kept a watchful eye on events in Geneva, headquarters of the wto. Cotonou includes protocols for sugar, beef and veal, and bananas; the rum protocol was discontinued. Cotonou's Banana Protocol, however, is significant for its lack of specificity. It voices continued support for ACP banana industries and commits the EU to continuing its efforts to make those industries more competitive. But it includes no promise of continued trade preferences and makes no reference to quotas for ACP bananas. The demise of preferences changed the basic nature of the system, reflecting the desires of the wto and paving the way for free trade in bananas.

8

The World Trade Organization
and the Banana Trade

Since its founding in 1995, the World Trade Organization has had a great impact on the banana trade, which suggests that it might affect the trade of other commodities as well. The WTO is a reflection of the neoliberal paradigm driving the world economy. Its creation was a response to the fourteen-fold increase in the volume of global trade between 1948 and 1994, an increase that proved beyond the ability of the earlier GATT to address. Technological advances in transportation, communications, and production systems contributed to the growth of trade and to its restructuring. According to Anderson (2000, 25), the growth permitted firms to disaggregate their production systems into spatially dispersed subprocesses that produced inputs and intermediate goods in different parts of the world. The new production systems increased the amount of international trade that occurred within firms and contributed to the growth of trade as a share of global output, from 7.7 percent in 1950 to 22 percent in 1999.

The WTO's primary goals include the reduction of obstacles to free trade and the development of a consensus on rules of international trade. These goals have generated great controversy, evidenced by the massive demonstrations outside WTO meetings in normally placid settings like Seattle. Many observers view the organization as an instrument used by transnational corporations to promote neoliberal policies that further globalization processes and yield results primarily beneficial to international capital.

This chapter focuses on the post–World War II evolution of the global trading regime, beginning with an overview of the GATT, from its creation as part of the Bretton Woods accords through its first four decades of operation. The discussion then turns to the Uruguay Round of talks, the GATT's final stage, which paved the way for the creation of the WTO. Finally, the WTO's involvement in agriculture, which ultimately shaped its role in the banana trade issue, is reviewed.

The GATT: Origins and Experience

The GATT is one of three primary institutions, along with the IMF and the World Bank, that grew out of meetings in Bretton Woods, New Hampshire, during World War II. Those meetings were intended to shape the postwar world economy, assist with the reconstruction of countries devastated by the war, and establish rules for international trade. The goal with regard to trade was to move toward reducing trade barriers and enhancing global economic integration. The GATT was not originally conceived to function the same way as the other two institutions. Rather, it was created in 1947 to provide the framework within which global trade would occur. The administration of the system was to have been entrusted to the International Trade Organization (ITO). That institution, however, was never created because the major trading parties declined to join. This denied the GATT an organizational structure, leaving it under-institutionalized. In the absence of strong treaty language, the GATT was forced to devise its own means of dealing with trade disputes and other issues involving international commerce. The result was a system that operated in an ad hoc fashion, which weakened its effectiveness (Jackson 2000, 68).

The dispute settlement arena manifested this weakness. Each time a problem arose, usually a complaint submitted by a member state, the GATT responded by creating a working party or, in later years, a dispute panel to consider the case. These temporary committees issued their decisions following a series of hearings, but the absence of a strong institution reduced the possibility of enforcing their decisions. It was easy for countries to ignore an unfavorable ruling, limiting the use of GATT's dispute settlement process by potential complainants.

The GATT's limited economic scope proved to be a second major weakness. The Bretton Woods system emphasized the economic interests and problems of the developed states of the capitalist world. The Soviet Union and its allies declined to join, and most LDCs were still ruled by European colonial powers when the accords were signed. The GATT reflects this first world bias in its focus on trade in manufactured goods; one of its major goals was to lower trade barriers for such products. For most of its history, the GATT system largely ignored trade issues involving primary sector commodities, including agricultural goods. Within the secondary sector, there was no GATT agreement covering textiles and clothing, two of the principal manufactured goods traded by LDCs. Thus, until recently the system was irrelevant to the concerns of most countries of the South, many of which had erected protectionist barriers for their limited secondary sectors and were principally exporting primary sector commodities. Relatively few LDCs joined the system. In reality, the GATT was not very inclusive until the 1990s.

The GATT's ad hoc nature was reflected in its evolution, which consisted of a series of "rounds," the term commonly applied to multiyear series of multilateral negotiations involving member states. The first round was the one completed in Geneva in 1947, culminating in the GATT itself. Four other early rounds, held from 1949 through 1961, focused on overcoming protectionism in developed countries and lowering secondary sector tariffs, but the number of participants in each round was rather limited (Carbaugh 2000, 194). Negotiations were generally tedious, approaching tariff reduction on a product-by-product basis.

The GATT's first truly significant round was the Kennedy Round of 1964–67. It was named for the U.S. president, who initially called for the round. The Kennedy Round marked a dramatic change in the GATT's approach because the product-by-product system was abandoned in favor of across-the-board negotiations on tariff reductions. The new streamlined process achieved a 35 percent reduction in tariffs on manufactured goods, lowering average remaining tariffs to just 10.3 percent of a product's base value.

Further changes were forthcoming during the Tokyo Round (1973–79), which lowered secondary sector tariffs further, to an average of just 4.7 percent. The true accomplishment of this round, however, was the initiation of a process of addressing the many non-tariff barriers to trade. Participating states agreed to codes of conduct in areas like import licensing, anti-dumping, customs valuation, government procurement, and technical barriers like product standards (Carbaugh 2000, 194).

The weaknesses of the system grew increasingly apparent as the global economy evolved. The rise of telecommunications and computer technologies added new dimensions to international trade as TNCs dispersed their operations. Many countries ignored the GATT and pursued bilateral trade arrangements. Trade-distorting subsidies beyond the GATT's purview were employed as a protectionist tool. These factors, and the absence of most LDCs, provided the impetus for a more comprehensive round of GATT negotiations.

The Uruguay Round: Transition to the WTO

The last of the GATT rounds of talks began in Punta del Este, Uruguay, in 1986 and continued until April 1994. The Uruguay Round was by far the most comprehensive and inclusive—in terms of both its economic scope and the number of participating states—of all GATT rounds. It addressed many of the system's weaknesses but was flexible enough to adjust to the changing geopolitical and technological milieu within which international trade occurs. The demise of the Soviet Union during the round eliminated the primary challenge to global capitalism and created a plethora of new economic opportunities. Furthermore, advances on many technological fronts were changing the nature of trade and stimulating its rapid growth. The GATT was ill suited to confronting these challenges; it was clear that new methods were needed. More than one hundred countries ultimately signed the Uruguay Round agreements that responded to these issues upon the conclusion of talks in Marrakech, Morocco (WTO Secretariat 1999, v).

The GATT accord signed in Marrakech was the first to include agriculture; it stipulated that developed countries reduce trade barriers in that sector and required them to complete a series of reduction commitments within six years after 1995. LDCs were granted a ten-year period to do so. The reduction commitments involved three areas. The first was market access: Developed countries committed to lowering their tariffs by 36 percent from 1986–88 levels, and LDCs committed to reducing theirs by 24 percent. The second commitment related to domestic support, essentially subsidies. Reductions of 20 percent were required, using the same reference period. Finally, government export subsidies for agricultural goods had to be reduced by 21 percent (CARICOM 1996, 36–37).

Despite the Uruguay Round's greater inclusiveness, several gaps remained. There was no explicit treatment of environmental matters related to global trade, an increasing concern among many critics who fear that rapid increases in trade volume occurred at the expense of environmental protection—especially in LDCs, where legislation and enforcement might not be very rigorous. Labor issues also remained beyond the purview of the new system; this was a major shortcoming, given that the industrial restructuring made possible by quaternary sector technologies had also changed the relationship between capital and labor. Such changes were evident in subcontracting arrangements often used by TNCs in the production of inputs in the preassembly stages of many industries. Labor practices of many subcontractor firms fell short of accepted standards in the home countries of the TNCs that sourced inputs from them. The freer trade facilitated by the Uruguay Round increased the ease with which TNCs could engage in such relationships.

After the signing of the new accords in 1994, many countries joined in the process. As of December 2005, the WTO had 149 members and 33 observers, with several applications pending—notably those of Russia, Saudi Arabia, and several small island states. When the GATT 1947 ceased operation at the end of 1994, it had 128 signatory states participating, with 33 of those having joined during the 1990s. Most new members were LDCs or former Warsaw Pact states.

One Organization, Many Agreements: The Structure of the WTO

The Uruguay Round yielded a more comprehensive series of agreements—sixteen altogether—and a more complex administrative structure, rendering the system difficult to understand.[1] The complexity was sufficient that it necessitated an intensive three-month training program to orient new WTO ambassadors to the WTO's institutional framework and plethora of agreements, accords, and schedules (Gomez 2001, personal communication).

Anderson (2000, 8) identifies the four key objectives of the WTO as (1) the development and enforcement of rules to govern international trade, (2) the provision of a forum in which member countries can negotiate and monitor the liberalization of global trade, (3) the improvement of trade policy transparency among members, and (4) the resolution of trade disputes among members. Whether or not one agrees that free trade is beneficial to most people, it would be difficult to argue against the need for rules to govern international trade or against the need for an organization to establish and enforce such rules. One of the purposes of GATT/WTO rules is to protect weaker, smaller states against discriminatory trade practices by larger, more powerful ones. Ostensibly, GATT's Articles 1 and 2 accomplish this. Article 1 establishes the most favored nation (MFN) principle of equal treatment of all member states by other members. Article 2 was designed to discourage the raising of trade barriers, thereby preventing larger states from exploiting their monopoly potential. It binds nations to tariff schedules that, once in place, cannot be increased. The presence of such rules also enables national governments to respond negatively to pressure by domestic interest groups to enact protective measures (Anderson 2000, 9).

The secretariat that loosely oversaw the GATT 1947 system became the permanent WTO Secretariat in 1995. The new body has stronger powers than its predecessor, particularly with regard to trade policy review and dispute settlement. The secretariat ensures that the WTO fulfills the five basic functions defined in Article 3 of the agreement creating the organization. Those functions include (1) implementing the multilateral trade agreements signed at the conclusion of the Uruguay Round, (2) the flexible implementation of those agreements where LDCs are concerned, (3) serving as a negotiating forum for future multilateral trade negotiations, (4) endeavoring to settle trade disputes among members, and (5) reviewing members' trade policies on a regular basis (WTO Secretariat 1999, 4). The WTO also assumed a controversial role in working cooperatively with the IMF and World Bank toward the goal of achieving "greater coherence in global economic policy making" (Anderson 2000, 9).

A Ministerial Conference occupies the top level of the WTO's administrative structure. Comprised of all member states, it meets irregularly but must do so at least once every two years to establish major policy directions. The Ministerial Conference meetings attract great public attention, as in Seattle (1999) and Cancún (2003). Below that top level, the General Council, also with all members involved, conducts the ongoing administration of the WTO. Usually, members' heads of mission to the WTO are also their countries' representatives on the General Council. The General Council also convenes as the Dispute Settlement Body (DSB) and the Trade Policy Review Body.

Below the General Council is a complex network of subsidiary bodies that focus on more specific aspects and sectors of trade, including goods, services, and intellectual property (WTO Secretariat 1999, 6–8).

The laborious task of reviewing national trade policies is oriented toward ensuring policy transparency among members. This work is entrusted to the secretariat's Trade Policy Review Mechanism (TPRM). The TPRM is not an enforcement arm of the WTO. It merely reviews members' trade policies to ascertain their compatibility with WTO rules, essentially performing an external audit of a nation's trade regulations. Problems discovered are dealt with by other branches of the organization. The reviews occur according to a schedule that reflects the country or unit's relative importance in global trade. The policies of the United States, Canada, the EU, and Japan are reviewed biannually. The next sixteen most important trading countries are reviewed every four years. For the smallest and poorest LDCs, reviews occur less frequently (Anderson 2000, 17). Siebert (2000, 142) suggests that regular policy review is necessary in an era of globalization to ensure that free trade enhances, rather than diminishes, competition. Regular reviews can help eliminate trade policies that give unfair advantages to domestic enterprises or home-based TNCs, allowing them to exploit their market power or create monopoly conditions.

The WTO implements the rules and policies contained in the Uruguay Round agreements, the broadest of which is the GATT 1994. A commonly held misconception about the WTO is that it replaced the GATT. Technically, this is untrue. The GATT still functions, although in a revised, expanded form known as GATT 1994. It continues to embody many of the principles of the original agreement (e.g., MFN). The major change is that the GATT now functions within a stronger organizational framework provided by the WTO Secretariat, which has responsibility for the oversight of the agreements signed in Marrakech.

The Marrakech accords include four sectoral agreements that extend the rules of trade into areas not addressed by the GATT 1947.

They cover an intriguing array of two traditional economic activities that have been a major part of world trade for centuries and two new sectors that reflect growing concerns of postindustrial economies. The Agreement on Agriculture and the Agreement on Textiles and Clothing bring two traditional sectors into the realm of global trade rules for the first time, while the General Agreement on Trade in Services (GATS) and the Agreement on Trade-Related Aspects of Intellectual Property Rights (TRIPS) do the same for two modern sectors that reflect the "tertiarization" of the post–World War II global economy. The GATS covers legal, financial, accounting, and software services, but not all countries signed on to the accord in all of those areas. For example, the United States did not sign on to the financial services agreement. Thus, when Japan did not grant the United States access to its financial services market, the United States could reciprocate by not opening its financial services markets to Japan. The TRIPS agreement provided protection lasting seven years in the case of trademarks, twenty years for patents, and up to fifty years for copyrights, thus favoring publishers, film studios, software developers, and pharmaceutical firms (Carbaugh 2000, 195).

The remaining agreements focus on mechanical aspects of trade such as the application of sanitary and phytosanitary measures (important in agriculture), subsidies, customs valuation procedures, rules of origin, import licensing procedures, preshipment inspection, and trade-related investment measures. These areas often lead to disputes. The Agreement on Import Licensing Procedures, for example, figures prominently in the banana trade issue because of the controversial nature of the EU import licensing system. Collectively, this array constitutes the primary body of international law governing the rules of trade.

The WTO and Agricultural Trade

The Agreement on Agriculture was a major, although controversial, change brought about by the Uruguay Round. Agricultural trade encountered many obstacles during the postwar period because of protectionist policies and high farm subsidy levels in the major market countries of the North. Agricultural goods are an important component of North-South trade and figure prominently in the export profiles of most LDCs. Their inclusion in the negotiations was an incentive to participate in a system that previously had been irrelevant to them. The controversy stems from the increasing activity levels of TNCs in the agricultural sectors of many LDCs, a trend that was well under way before the Uruguay Round but that was stimulated by the promise of increasingly free trade in agricultural commodities. This trend interacts with other processes, including the export imperative affecting many LDCs struggling to meet payments on their sizable foreign debts.

Agriculture's inclusion in the new GATT has subjected it to WTO rules and requirements, with both positive and negative consequences. It yields new opportunities for countries in the South through the disassembling of protectionist policies and the reduction of subsidies in first world market countries. But it also affects special trading relationships that are based on historic patterns and linkages, including the Lomé Convention, because the new rules determine them to be impediments to free trade.

Much of the impetus for including agriculture in the new agreements came from the South, particularly from the Cairns Group, which is composed of eighteen major agricultural exporters, mostly LDCs.[2] Those countries wanted to ensure that any agreement reached during the Uruguay Round extended the process of trade liberalization into the agricultural sector. They were particularly intent that the round address the issues of domestic farm subsidies, agricultural tariff reduction, and export subsidies (Australian Department of Foreign Affairs and Trade 2000). Unlike secondary sector subsidies, which declined under the GATT 1947, agricultural subsidies in Organization for Economic Cooperation and Development (OECD) states remained high and were accompanied by other protective measures as well. According to Anderson (2000, 12), rich countries could afford the "distortionary costs" of protecting their agricultural sectors from foreign competitors. In addition, long-term protection of inefficient agricultural production contributed to global overproduction and lower prices for many products. GATT 1947 fostered this situation through accession procedures and waivers that allowed members to continue past protectionist practices by listing such products in their schedules.

The Marrakech Protocol's Agreement on Agriculture obligates all member states to reform their agricultural sectors and pursue market-oriented policies by addressing three major areas. The first is market access, where the agreement's goals include "tariffication" (the creation of a tariff-only regime as the sole permissible means of protection in agriculture), the reduction of tariffs over time, and the binding of the remaining tariffs. Strict, complex guidelines apply to any exceptions to this policy, although they are applied more leniently to select LDCs. Overall, this policy is a dramatic change from previous norms on agricultural protection (WTO Secretariat 1999, 54).

Domestic support is the second area addressed, with the goal being to induce policy shifts within countries that will reduce distortions in agriculture. Policies such as stockholding for food security, disaster relief, structural adjustment programs, domestic food aid for the poor, and government support for agricultural research are classified as "green box" measures that do not cause distortions, and they continue to be permitted. "Amber box" measures, those that distort trade, are discouraged. Government purchasing of agricultural produce at guaranteed prices falls into this category. Members must commit themselves to reducing such measures through the programmed reduction commitments included in their Marrakech Protocol schedules (WTO Secretariat 1999, 56).

Export subsidies are the third area addressed by the agriculture agreement. Unlike GATT 1947, GATT 1994 bans such subsidies but does so with specific exceptions. The exceptions include export subsidies subject to reduction commitments, as specified in the member's Marrakech Protocol schedule of commitments, and subsidies by

LDCs, which are consistent with the agreement's provision for special and differential treatment (WTO Secretariat 1999, 59). The Marrakech Protocol permits exceptions to trade rules where there are concerns about sanitary conditions, where mandated agricultural reforms might have a negative impact in selected LDCs, and in countries with high dependence on food imports.

With its foray into agriculture, the Marrakech Protocol clearly broke new ground, but not everyone has interpreted it in the same manner. Subsidies remain in place in many countries; other countries are frustrated by what they feel is a lack of progress in the sector. The EU is continually criticized in this regard. The WTO rules are contradictory to other EU priorities, notably efforts to preserve Europe's rural sector despite high labor costs and the lack of competitiveness of many of its farmers. To some observers, the EU's interpretation of the WTO agricultural subsidy policy is not that subsidies should decrease but that they should increase at slower rates (Suescum 2001, personal communication). The WTO's interest in the primary sector also interjects the GATT in unprecedented ways into the relationship between the EU and the ACP states.

The Lomé/Cotonou System: Adjusting to the WTO

The Lomé/Cotonou system was significantly affected by the transition from the GATT 1947 to the WTO. The greater emphasis on free trade and the broader scope of the latter brings it more directly into the trade relationships embodied in the Lomé Conventions. The mere threat of WTO intervention affected the negotiations for the Cotonou Agreement before 2000 and manifested itself clearly when the EU applied for a WTO waiver for Cotonou's temporary trade preferences for ACP states. The trade preferences violate Article 1 of the WTO agreement, the article providing for MFN treatment. They allow the EU to continue to provide duty-free access to a variety of goods from the ACP group without reciprocity. Under the old system, the European Community easily obtained waivers for the first four Lomé Conventions. In 1995 it again succeeded in obtaining a waiver for the second phase of Lomé IV from the newly formed WTO. However, the Cotonou waiver process was complicated by the ongoing banana trade dispute and the reticence of Latin American banana exporters to support a waiver until the dispute had been satisfactorily resolved. There were also concerns about the lack of specificity in Cotonou's Banana Protocol. Rather than help resolve the dispute, the Banana Protocol seemed to allow for reopening the preferences issue after 2006 if ACP countries had not successfully upgraded their banana industries by then (Suescum 2001, personal communication).

The need for waivers is derived from the concept of obligation, which is an integral part of the GATT/WTO system. Each country, in the act of signing WTO agreements, assumes a variety of obligations with regard to upholding the rules of trade specified within those agreements. The rules are based on fairness and equal treatment of fellow members, and they discourage trade policies that do not afford reasonably equal access to markets. Accords like Lomé and Cotonou, which occur outside of this framework and contain arrangements that do not conform to the WTO rules, constitute violations in the absence of a waiver. While the first four Lomé Conventions did provide for preferences for the ACP group, those preferences mostly extended to primary sector commodities that were unaffected by GATT 1947 rules. Those waivers were uncontroversial and many participant countries had not signed the GATT. This changed after the Uruguay Round. Many ACP goods receiving EU preferences are covered by GATT 1994 rules, and many countries that could be affected negatively by those preferences are now WTO members.

The WTO's waiver-granting procedures were tightened up and more clearly defined by the GATT 1994 agreements. While the GATT 1947 did little more than state voting requirements for the granting of waivers, the 1994 version goes considerably further. New waiver procedures appear in the general WTO agreement and in a special section of the agreement entitled "Understanding in Respect of

Waivers of Obligations under the General Agreement on Tariffs and Trade 1994." Members wishing to be relieved of particular obligations imposed on them by the GATT now must explain what trade measures they propose to implement and define their policy objectives when requesting a waiver. They must also provide reasons why those objectives cannot be achieved through measures corresponding to GATT 1994 regulations. The WTO General Council is responsible for granting waivers; the voting requirements to approve them are outlined in Article 9 of the WTO agreement. The rules also require that specific termination dates be attached to any waiver granted. Finally, all waivers that existed as of January 1, 1995, expired or were terminated by January 1, 1997, unless they were extended under the new rules. Thus, by 1997 all waivers were governed by GATT 1994 regulations (WTO Secretariat 1999, 48–49).

In early 2001 the EU's Cotonou waiver application was still pending. Because the Cotonou Agreement went into effect in March 2000, the EU technically was in violation of WTO regulations. The previous waiver expired in February 2000 along with Lomé IV (Bouflet 2001, personal communication). In the past, waiver requests were seldom put to a vote; they usually were approved by a qualified majority (under GATT 1947) or by consensus (under the WTO). Today, however, a 75 percent vote in favor is needed for approval. A favorable vote for the Cotonou waiver was far from certain (Bouflet, Brown, and Suescum, all 2001, personal communication). Although it eventually did occur, it required the resolution of the banana dispute case.

Dispute Settlement in the WTO: Breaking New Ground

Improving the trade dispute settlement process was a primary goal of the newly constituted WTO system. GATT 1947 did not contain detailed language on dispute settlement, and the process evolved in an ad hoc manner that was underused by member states, reducing its potential for a positive impact. According to Anderson (2000, 18), there were only three hundred cases brought before GATT 1947's dispute settlement process during its forty-eight-year history. This average of six cases per year indicates members' lack of faith in the process, a situation usually attributed to the GATT's weak enforcement powers.

The WTO agreement includes stronger, more detailed dispute settlement language and applies that language to the GATS, the TRIPS, and the GATT 1994. Its broader scope opens up the possibility that compensation awarded in a settlement for harm done in one sector could take the form of sanctions in a different economic sector. The EU's Zelie Appleton (1996, personal communication) labeled the system "more precise" and noted that it would no longer be possible for one member or entity to unilaterally block adoption of WTO dispute panel decisions (Appleton 1996, personal communication). The WTO system also addresses time concerns, making the process more efficient through the imposition of deadlines for each stage of a complaint settlement proceeding (WTO Secretariat 1999, 17). The result has been an increase in the number of cases, with 168 during the first fifty-two months of the system's existence (Petersmann 1999, 3). Some view this as an indication of acceptance of the new procedures, while others see it as a desire to test them or to challenge ambiguity in the Uruguay Round agreements. Jackson (2000, 69) suggests that both are true. One-third of the cases were initiated by the two major trading powers: the United States (thirty-five cases) and the EU (twenty-one cases). Both countries have lost cases and accepted the outcome, a departure from pre-1995 behaviors (Siebert 2000, 156). Nearly two-thirds of the cases were complaints against developed states. Furthermore, rulings made by the panels have resulted in policy changes in the countries that were found to be in violation of the WTO rules (Anderson 2000, 18).

The new system did extend some core aspects of the old process, including the right of a member to consultations with other members and the principle of nullification and impairment. The latter can be invoked when a member state feels that it was deprived of benefits because of the trade policies of another member state (WTO

Secretariat 1999, 18). The basic principles and objectives of the new system are quite clearly identified, with the preferred outcome of a dispute being the withdrawal of the policy that is in violation of wto rules. Failing that, compensation to the harmed party should occur. Retaliation in the form of sanctions is the least desirable outcome and is to be used only when the first two results do not occur and only with wto authorization. Members are required to use the dispute settlement process to settle differences; unilateral action is prohibited (wto Secretariat 1999, 20).

Institutionalization also differentiates the new system from its predecessor. Its primary institution is the DSB, essentially the wto General Council but with a different chairperson for the purposes of dispute settlement. The DSB has sole authority to establish a panel to hear a dispute. Unlike the old system, the new system forbids a member from refusing to participate in the process if a complaint is filed against it and the DSB creates a panel to hear the complaint. The DSB's decisions are reached by consensus, in keeping with the wto's general decision-making approach (Jackson 2000, 77).

The dispute panel is the second institution. A panel is set up to hear a specific complaint and is disbanded once the matter is resolved. Dispute panels include either three or five members who are professionals drawn from countries not involved in the dispute. The third institution is the Appellate Body, which had no equivalent under the GATT system. Once a panel has rendered its decision, the parties involved in the dispute have the right to appeal. The Appellate Body has seven members, any three of whom can be convened to handle a given case. The Appellate Body's decision is sent back to the DSB, where, if approved, it must be unconditionally accepted by the disputing parties (wto Secretariat 1999, 20–22). All of these institutions were called into play during the prolonged banana trade case.

When a member is found to be at fault through the dispute settlement process, the member has thirty days to inform the DSB of the actions it will take to rectify the matter and comply with the decision. In extraordinary circumstances, "a reasonable amount of time" beyond those thirty days will be granted to the party to bring its policies into line. Once that time has elapsed, if no solution has been proposed, the harmed party may request authorization to retaliate. Preference is given to retaliatory actions within the same sector, although approval may be granted for retaliatory measures in other sectors (WTO Secretariat 1999, 24).

Timetables were established in the WTO agreement for each stage of the dispute settlement process, which is an indication of the WTO's will to address issues quickly. The timetables are considered one of the keys to the success of the new system, although even with specific timetables, the granting of extensions can slow the process and lead to frustration. Such was the case with the banana dispute. No more than nine months should elapse between the establishment of a dispute panel and the approval of its report by the DSB. If a panel report is appealed, the nine-month period can be extended to twelve months to allow time for the appeal process, and the DSB must adopt or reject an Appellate Body report within thirty days of receiving it (WTO Secretariat 1999, 27–30). Despite the fact that more than a year can elapse between the initial request for consultations and a resolution, this pace of events is quite streamlined in comparison with the previous process (Anderson 2000, 18).

A corollary issue related to dispute settlement in the WTO is who is permitted to participate in the complaint process. Under GATT 1947, only initial negotiators and principal suppliers of a good were granted the right to compensation for losses of benefits suffered from tariff bindings on that product. The initial negotiator is the country that first obtained a tariff binding for a product through bilateral negotiations, and the principal suppliers are countries that now export more of the product than the initial negotiator, normally with a threshold of 10 percent applied. The GATT 1994 extends lesser rights to countries with a substantial interest in the trade of the good but who do not fit into either of those categories. According to the WTO Secretariat (1999, 49): "The Uruguay Round . . . has modified these rules to recognize that, for some supplying countries, a small share in the market for the product concerned may in fact be of great economic importance. The change should benefit the smaller WTO members and particularly developing countries." This ostensibly permits smaller countries the right to negotiate for compensation even if their total market share falls below that of principal suppliers and the initial negotiator. To qualify for this inclusion, a country must demonstrate that its exports of the good represent a significant percentage of its total export profile.

Despite this provision, some suggest that the Caribbean suffered with the transition from GATT 1947 to the WTO (Brown 2001, personal communication). Their problem was exacerbated by the fact that the transition occurred on the heels of the EU's adoption of the SEM. Before 1993 eastern Caribbean countries met the 10 percent threshold in the UK banana market, so they qualified as substantial suppliers. After the creation of the SEM, the WTO considered the entire EU as one market within which each exporting country must achieve the 10 percent substantial suppliers share. Caribbean countries cannot do so, relegating them to secondary status. Despite the fact that several Caribbean states can demonstrate the importance of bananas to their total export profile, this proof has been difficult to parlay into additional clout within the WTO.

9

The U.S.–EU Banana
War Heats Up

The reactions to the European Union's implementation of Council Regulation 404/93 in July 1993 ensured that the policy would be tested in the international arena. The economic importance of the banana trade and the complexity of the relationships it engenders increased the vigilance of all of the parties involved as they monitored one another's efforts to preserve or alter the EU policy. The so-called banana war that resulted from these machinations quickly gained notice in the European, Latin American, and Caribbean presses before finally making headlines in North America in 1999. The major issues were trade preferences, quotas, import licensing procedures, and tariffs; the various battles of the "war" showed the lengths each side went to to defend its interests in this important industry.

The banana war began with the battle lines clearly drawn. On one side were the EU and its ACP partners, defending trade preferences, higher tariffs, and import licenses. Ironically, the EU is normally one of the primary motors of globalization, but bananas proved to be an exception to this rule. On the other side, a theoretically unified

Latin American front promoted free trade in bananas, essentially a pro-globalization position. The United States initially seemed uninterested in the issue, at least at the government level. But that situation did not continue for long.

The Initial GATT Challenges

In 1993, shortly after the implementation of the EU's new banana regime, Colombia, Costa Rica, Guatemala, Nicaragua, and Venezuela filed complaints against the new policy with the GATT.[1] Hearings were held and the EU based its defense of ACP banana preferences on its Lomé waiver (Hirsch 1998, 202). The dispute panel ruled in January 1994 that the EU policy was unfair to third-country producers, and it condemned the ACP tariff preference as discriminatory. The panel could not justify the EU's extension of tariff preferences to ACP states that were not GATT signatories while GATT members were excluded from such benefits. The ruling also denounced the size of the tariff on third-country bananas for exceeding the EU's normal 20 percent tariff on agricultural goods (USTR 1994, 1). It also faulted the Lomé system for its lack of reciprocity for European goods (European Commission 1994, 11). In addition, the panel addressed the import licensing system, determining that because the incentive system used in granting category B licenses ensured favorable treatment to fruit from EU and ACP sources, it violated GATT's MFN principle and its requirements on national treatment (USTR 1994, 2). The panel did not rule that import licenses themselves were illegal, but it said that their use must occur within a GATT-compatible framework. Thus, the panel ordered the EU to redesign its policy to be compliant with GATT rules.

Latin America was encouraged by the decision, and the major exporting countries anticipated greater access to the EU market. ACP states, on the other hand, were worried that the decision threatened the viability of one of their most important industries. Attention there focused on obtaining a GATT waiver for the second stage of the Lomé IV Convention (*New Chronicle* 1994b, 1). The EU response was mixed: free trade proponents such as Germany welcomed it, but France and the UK viewed it as a defeat. What really mattered, however, was the response from the European Commission, the EU administrative body that would ultimately need to make any alterations to the policy.

The Banana Framework Agreement

The official EU response to the GATT ruling was to ignore it. In doing so, it illustrated three important dimensions of the banana dispute. First, it exposed the weaknesses of the GATT dispute settlement system that would eventually be addressed by the WTO. The EU could ignore the ruling because the GATT had no enforcement authority. Second, EU actions after the ruling exposed the lack of Latin American unity on the banana trade issue when it redrew the battle lines of the banana war with the Banana Framework Agreement (BFA). Third, the failure of the BFA to settle the issue led to the involvement of the United States, indicating that the EU had misread American interests in the dispute. Where the BFA focused on the quantity of third-country bananas, U.S. interests revolved more around the banana import licenses and their impact on the earnings of its three large TNCs.

The Banana Framework Agreement was negotiated between the EU and the GATT plaintiffs. Costa Rica, Colombia, Nicaragua, and Venezuela all signed the accord, but Guatemala refused to do so (European Parliament 1998, 12). The signers agreed not to challenge Council Regulation 404/93 for the remainder of its lifetime (European Commission 1994, 12). The BFA took effect in April 1995. It addressed several aspects of the GATT complaint, particularly EU market access, by providing each of the four countries with its own banana export quota to the EU within the overall tariff-rate quota for third-country bananas (table 6). The BFA ended the free market within the third-country tariff-rate quota. Instead, it yielded a situ-

TABLE 6: Annual share of Banana Framework Agreement quotas

Country	Annual share of tariff-rate quota (%)
Colombia	23.03
Costa Rica	25.61
Ecuador	26.17
Panama	15.76
Others	9.43

Source: FAO 1999

ation similar to that governing the import of ACP bananas, ironic given that Latin America had argued for a free market for all bananas entering the EU.

The BFA also permitted exporting countries to issue export certificates for bananas. Their governments could charge fees for the certificates to generate public-sector funds for use in supporting the industry, an important provision in gaining their support for the agreement. According to Ronald Saborio Soto, export certificates were included in the BFA because Costa Rica and Colombia wanted some of the rents generated by the EU import policy to accrue on their side of the Atlantic (Saborio 2001, personal communication). Before the BFA, European firms realized all the rents through the sale of import licenses, often at prices higher than those for the bananas themselves. All Latin American countries experienced declines in public-sector revenues from bananas during the early stages of the EU policy when they lowered export taxes to gain short-term advantages within the third-country tariff-rate quota. The BFA represented a means of recovering some of those losses.

Under the BFA, Costa Rica and Colombia required export certificates only for bananas destined for the EU market. Costa Rica assigned certificates by farm, based on historical production records. Certificates were distributed free of charge to all farms. The farmer could then add the certificates to the cost of the bananas charged to the export company, inevitably a foreign TNC, at a price of two dollars per box (Saborio 2001, personal communication). This charge lowered by about two dollars per box the cost of the import licenses paid by the company on the other side of the Atlantic. So the cost of the bananas did not increase as a result of the export certificates, but the distribution of the rents changed.

In negotiating the BFA, the EU was able to remove four countries from the list of those opposing its banana importation policy. It destroyed Latin American unity on the issue and prolonged the dispute. The BFA was concluded in time for the EU to list it in its Uruguay Round schedule of conditions under which it would accede to the newly created WTO. When, in 1996, the BFA itself was challenged in WTO complaints filed by the United States and non-BFA Latin American states, the EU argued that the BFA's listing in its schedule "immunized it from the challenge" (Hirsch 1998, 203). The dispute panel ruling on the complaints banned the certificates but preserved BFA country quotas, although the panel ruled that all four major third-country suppliers of bananas to the EU must agree to the quota system. Because Ecuador and Panama opposed the use of quotas, the 1997 WTO decision effectively subverted the BFA system and thwarted the EU initiative.

A Surprising Development in the Eastern Caribbean

Challenges to the EU banana regime were viewed as threats in the eastern Caribbean, despite the policy's country quotas for ACP bananas. This view was held because of the actions of major TNCs like Chiquita, which made overtures to the region's farmers, particularly in St. Lucia, where it met with the Banana Salvation Committee opposed to government banana policies. The meetings yielded no results, but the mere presence of Chiquita in the region generated suspicions that its real agenda was to destabilize the eastern Caribbean's industry to remove it as a challenger in EU markets (Mark 1997, personal communication).

A more serious situation was generated by Geest's 1994 decision to leave the industry, which resulted from financial problems from Tropical Storm Debbie and an outbreak of Sigatoka negra on its Costa Rica plantation. Company earnings dropped, and compensation from the UK government did not offset the decline. Rumors circulated in the eastern Caribbean that several TNCs, Chiquita among them, were interested in purchasing the company, whose ACP activities entitled it to valuable import licenses in the EU. There was concern in the Windward Islands that such a purchase would force the islands to deal with a company that would not be sensitive to their more general needs (Brinard 1996, personal communication).

Geest began negotiations with Noboa, an Ecuadorian firm, for the sale of its banana division. The talks failed but sent shock waves through the region, where many feared that Noboa would favor Ecuadorian bananas over eastern Caribbean bananas in the European market and have little incentive to maintain Geest's historic shipping links, whose importance to the islands' economies extended beyond bananas. These fears stimulated the region's governments to act. With a high level of cooperation, they established a consortium with Fyffes, the other major TNC marketing bananas in the UK, which was perceived as more likely to support the future of the Windward Islands banana industry. In 1995 the consortium purchased Geest's banana division for 147.5 million British pounds, forming the Joint Ventures Company, a firm in which Fyffes became the minority shareholder.[2] The sale included Geest's losing venture in Costa Rica, two ships totaling 640,000 cubic feet of cargo space, and, significantly, Geest's back haul business, which transported 85 percent of the eastern Caribbean's imported goods from the UK (Sandiford 2000, 52).

The buyout led to important changes in the institutional framework of the eastern Caribbean industry. WINBAN was abolished and replaced by the Windward Islands Banana Development Corporation Ltd. (WIBDECO), an umbrella entity that covered all four countries and purchased Windward bananas for UK and EU markets. Where

WINBAN was a loose parastatal association, WIBDECO was a profit-oriented business and functioned as such. It operated separately from the Joint Ventures Company, which marketed Windward bananas in Europe. WIBDECO also carried on most of the day-to-day mechanisms that were put in place under WINBAN, providing a degree of continuity to the system.

In 1999 WIBDECO encountered a major challenge. A change in political direction in St. Lucia, the largest Windward banana producer, led to privatization of the banana industry there and the breakup of the SLBGA into four smaller organizations. WIBDECO's monopoly as the only party entitled to purchase the farmers' produce ended when the largest of those four companies—the St. Lucia Banana Corporation—chose to leave the WIBDECO–Joint Ventures umbrella and find another means of selling its bananas (Reid 2000, personal communication).

The Geest purchase was welcomed with both enthusiasm and anxiety, but it had the full commitment of the four countries involved. Cooperation among them had always been high, but the new arrangement imposed a higher degree of risk directly on Windward governments. There was a sense that the venture must succeed, that failure would spell disaster given the deep roots of banana farming in the four societies. According to Durand (1996, personal communication), the buyout was essential to the industry's ability to overcome Geest's major shortcoming—its lack of ownership in the banana industries that supplied it. Many viewed Geest as more responsive to its shareholders than to the needs of the Windwards' farmers. The new consortium involved governments of democratic countries in which banana farmers were a significant voting block. It gave the islands more control over the industry, allowing them to work more effectively to protect its well-being. In addition, increased control over shipping arrangements allowed the islands to arrange transportation for other commodities as well, possibly to the point where shipping itself would become a profit center for the new company (Gregoire and Fadelle 1996, personal communication). The new system had considerable support among the region's farmers, who were hopeful that it would pay better prices for their fruit and provide more investment funds.

The United States Joins the Fray

The United States refrained from participation in the early stages of the banana war between Latin America and Europe. Given the United States' history of intervention in Latin America on behalf of its corporations, its silence on the EU banana importation policy created confusion and consternation in the region (Zúñiga 1994, personal communication). The lack of response was likely attributable to the fact that the three U.S.-based TNCs tried to adjust to the EU policy by acquiring category B licenses through mergers, buyouts, or joint ventures with EU fruit firms to maintain a strong presence in the EU market. Dole and Del Monte were quite successful in doing so, but Chiquita was not. As Chiquita's market share declined, it turned to political pressure in Washington, and the United States ended its silence.

On January 9, 1995, U.S. Trade Representative Michael Kantor advised the European Commission of his belief that the EU banana regime was "adversely affecting U.S. economic interests" (Kantor 1995). He warned that this would lead to retaliatory measures if the EU would not negotiate the matter in a "spirit of practical compromise." Sir Leon Brittan, the EC commissioner to whom Kantor's letter was addressed, reaffirmed the EU belief that its banana policy was WTO-compliant. He reminded Kantor that "no withdrawal of concessions, whether in goods or services, is permitted except following full WTO dispute settlement procedures" and advised him that "Europe clearly cannot accept any unilateral action against our interests in breach of those obligations" (Brittan 1995, 1). This exchange defined the direction of the banana war for the remainder of the decade. The United States positioned itself among the parties experiencing loss as a result of the EU policy, and the EU saw the U.S. position as a thinly disguised effort to support its TNC. Many EU officials believed that Chiquita was the force driving U.S. banana policy and that it used political contributions to do so. This opinion appears in EU files; it does not appear in public documents, although Brittan stated: "The whole process is driven by politics in the United States. It is driven by the fact that Chiquita is a company that gives money to the political parties . . . that the president of Chiquita is very close to Senator Trent Lott" (BBC News 1999a, 2). Chiquita did file a complaint against the EU policy under Section 301 of the U.S. trade law in 1994, before Brittan's statement (Sandiford 2000, 4). The European press picked up the political tone as well, presenting the issue as one of Europe defending the survival of small-scale Caribbean and African farmers against efforts by U.S.-backed transnationals to destroy them. Most European publics also shared this perspective.

The Caribbean ACP group was caught between its two giant associates. Linked to the EU through Lomé, it also had very important relations with the United States, its primary source of tourists and the major geopolitical force in the region. Caribbean states realized they had few bargaining chips at their disposal. They carried little weight within global banana circles, compared to Latin America, the United States, or the EU. Therefore, they had to use whatever weapons they could command on the battlefields of the banana war. One potentially effective weapon was what many suggested was a possible replacement activity for bananas, should the dispute end with an unfavorable decision for the Caribbean: drugs. They argued that if small-scale Caribbean farmers could not export bananas, many would turn to illegal crops like marijuana (Mark 1997, personal communication). Given the effort and sums of money that the United States had spent on its war on drugs and the geographic reality of the Caribbean as a transshipment zone, many saw this threat as a means of gaining U.S. attention for their plight. Migration, especially illegal immigration, was a second weapon in the arsenal. A sharp economic downturn in the Caribbean as a result of the collapse of the banana industry would likely yield, as it had in the past, increased flows of undocumented workers to North America. The U.S. response to these threats was to appear willing to negotiate, but it also persisted with its wto case.

wto Complaints: The Battle Heats Up

In February 1996 the United States, Ecuador, Guatemala, Honduras, and Mexico filed complaints to the wto against the EU banana importation policy (McMahon 1998, 104). The wto responded, authorizing the establishment of a banana panel through the DSB and setting the dispute settlement process in motion. Lasting from May 1996 through late September 1997, the banana case provided a test of the new system. In his analysis, Hirsch (1998, 201) describes the case as the "most complex" yet encountered by the wto, as it dealt with several key articles of the GATT 1994 and the GATS. It penetrated to the very core of the GATT/wto system, contesting the MFN principle itself. The precedent-setting case also established that a policy could be covered simultaneously by provisions of both the GATT and the GATS (Hirsch 1998, 202).

Initially, the EU tried to deflect the challenge by charging that the United States and Ecuador should not be permitted to file complaints in the case. It argued that the United States did not export bananas and that Ecuador did not list bananas in its wto accession protocol schedule when it joined in 1995. The United States replied that the European subsidiaries of its fruit TNCs experienced losses in their service sectors due to discriminatory licensing provisions of the EU policy. The banana panel upheld this link between the service and production sectors of the industry when it announced its findings in May 1997. It also ruled that being a banana exporter was not a prerequisite for filing a complaint and that the EU failed to prove its licensing system was not harmful to the subsidiaries (Hirsch 1998, 206, 221). The ruling had wide potential impact because it set a precedent that allowed countries to represent the interests of their TNCs abroad within the wto dispute process.

The panel also decided in Ecuador's favor, ruling that Ecuador's failure to list bananas in its accession protocol did not preclude it from filing complaints based on GATT 1994 articles related to the case. The panel also determined that Ecuador was not obligated to accept the validity of the BFA as contained in the EU's schedule (WTO 2000, 1077).

With regard to the complaints themselves, most rulings were unfavorable to the EU. The DSB determined that bananas were like products for the purposes of the various GATT articles reflected in the complaints, whether they come from EU, ACP, BFA, or other third-country sources (WTO 2000, paragraph 7.63). This deflected the EU claim that its import tariff-rate quotas dealt with different products and were separate, and thus were not discriminatory. The panel also ruled that the EU had only one banana importation system, not three. The issue, then, was to determine whether or not that system—the EU's allocation of TRQ shares—was similarly applied to all affected WTO members, in accordance with WTO rules (WTO 2000, paragraph 7.82). WTO rules prohibited application of different regimes to different members, except where all suppliers with a substantial interest in the good agree to such an arrangement (Hirsch 1998, 208). In addition, the allocation of quotas to BFA and ACP countries not having a substantial interest in supplying bananas to the EU (Nicaragua, Venezuela, and several ACP states) but not to Guatemala, which did have a substantial interest, was ruled inconsistent with GATT 1994 (WTO 2000, paragraph 7.90).

The Lomé system also came under scrutiny, with mixed results. The EU based its case on an interpretation that its Lomé waiver applied through the expiration of the Lomé Convention in 2000, allowing it to grant preferential access to the ACP exporters (McMahon 1998, 105). The plaintiffs argued for a strict interpretation of the waiver that would permit the EU to maintain the pre-1993 import levels but not to enhance them, as the banana regime appeared to do. The banana panel ruled that it was not unreasonable for the EU to conclude that Lomé required it to allocate country-specific TRQ shares to traditional ACP banana suppliers equal to their pre-1991 best-ever exports to the European Community, but the ruling stated that the EU was not required to allocate TRQ shares in excess of a country's pre-1991 best-ever exports (WTO 2000, paragraph 7.103). The panel also ruled that the waiver did not cover creation of separate licensing categories because the categories did not exist when the waiver was granted (Hirsch 1998, 211).

The EU banana import license system was a major target of complaints, with problematic results for the EU. The DSB ruled that the WTO licensing agreement applied to tariff-rate quota licensing procedures (WTO 2000, paragraph 7.156), and it found no legal basis for an a priori exclusion of the EU banana import licensing regime from GATS rules (WTO 2000, paragraph 7.286). Also, it determined that licensing procedures for traditional ACP bananas, third-country bananas, and nontraditional ACP bananas should be examined as one licensing regime (WTO 2000, paragraph 7.167). The allocation of 30 percent of third-country and nontraditional ACP licenses to category B operators was ruled a violation of the WTO requirement of equal opportunity to import goods regardless of source (WTO 2000, paragraph 7.182). The panel determined that the licensing system offered unfair benefits to ACP operators and placed unfair burdens on third-country and nontraditional ACP operators (Hirsch 1998, 211). Finally, the DSB found that the matching of EU import licenses with BFA export certificates was a violation of WTO rules (WTO 2000, paragraph 7.241).

Overall, the DSB findings created problems for the EU, and in June 1997 it announced its intention to appeal the case. This activated the WTO's Appellate Body, which reviewed the DSB report and heard additional arguments before rendering its final decision in September 1997. For the most part, it supported the banana panel's decisions, finding the EU policy inconsistent with the MFN principle and various GATT, GATS, and licensing agreement rules. It upheld the right of the United States and Ecuador to file complaints in the case. The Appellate Body actually worsened matters for the EU, strengthening the DSB decision by ruling that, in the absence of an agreement among all parties, a quota could only be assigned based on best-year amounts, and this could occur only if the best year fell within a specific reference period that was agreed to by all parties (WTO 2000, 691, pt. 255j). This, of course, had not occurred.

The Appellate Body instructed the EU to redesign its banana importation regime to be WTO-compliant, giving it until January 1, 1999, to do so or face possible sanctions from the parties experiencing losses from the noncompliant policy. Unlike the 1994 GATT decision, the WTO's mandate was one the EU could not ignore, so in late 1997 it began revising its banana regime, moving the banana war into its next stage.

The EU Tries Again

After the DSB and Appellate Body findings were announced, attention shifted back to the EU, which began work on a successor arrangement—that is, a policy to replace Council Regulation 404/93. The European Commission functionaries responsible for devising a new regime consulted extensively with the other parties involved, focusing on the aspects of the old regime that were ruled incompatible with WTO regulations—notably, individual country quotas and the licensing system.

ACP country quotas continued to prove controversial. They were staunchly defended by the twelve traditional ACP banana states. In 1998, however, the EU informed its ACP partners that individual country quotas would be eliminated in favor of a WTO-compatible global ACP quota. That quota would remain at the same level as the total of the twelve country quotas; individual states would have to work to become competitive within the quota, which was a primary EU objective for ACP producers. This change undermined ACP unity, which had been strong until that point. Belize's trade negotiator, Richard Reid, saw this as an EU effort to compartmentalize the ACP states into regional blocks (Reid 1998, personal communication). The proposed change was not viewed favorably by the Windward

Islands and Jamaica, which had experienced export shortfalls during the 1990s but were confident of their ability to fill their quotas in the future. A global quota, however, would allow Belize to expand its industry, generate economies of scale, and become more competitive. Belize saw the global quota as its opportunity to increase exports to the EU to a level that would allow it to have its own ship, one that did not also stop in other countries (Flores 1998, personal communication).

In October 1998 the EU's Council of Ministers approved a new banana regime, Council Regulation (EC) 1637/98, to replace Council Regulation (EEC) 404/93 as of January 1, 1999. It was remarkable for how little it changed the 1993 policy. Council Regulation 1637/98's changes included a tariff preference of two hundred euros per ton for nontraditional ACP bananas over third-country imports. It also established three quotas and maintained the two earlier quotas for third-country bananas, for a total of 2.53 million tons.[3] The third of the three new quotas was a global ACP quota, slightly reduced to 850,000 tons. The first two new quotas were allocated, in the absence of agreement among all major suppliers of bananas to the EU, to states with a substantial interest in the supply. The new policy also gave the right to import bananas to operators who effectively imported bananas during the 1994–96 reference period, excluding operators who traded in licenses but did not actually import bananas. Those with licenses could import from any country in the quota group. Thus, WIBDECO could gain EU licenses to import fruit from Latin America (Sandiford 2000, 67). Finally, it transferred the right to allocate import licenses for the three newest EU members from Germany to authorities in each of the three states.

The reaction to Council Regulation 1637/98 was mixed. The Caribbean Banana Exporters Association (CBEA) felt that the policy was less advantageous for the Caribbean due to the elimination of ACP country quotas. Eliminating the country quotas reduced market access security and decreased the incentives for operators to continue to purchase Caribbean bananas, with their higher production costs (CBEA 1999, 2).

The complainants rejected the new EU proposal, feeling that it did not fundamentally alter the licensing system related to category B licenses, the 30 percent of the licenses for dollar-zone bananas allocated to EU importers that traditionally handle ACP bananas. They also objected to the direct allocation of category B licenses by authorities in the three new EU members to importers in those countries. Previously, those licenses went through Germany, which vocally opposed the EU banana policy. The change brought criticism from the Federal Association of German Fruit Trading Companies, which complained that the new regulation would not halt the trade in licenses. The licenses were valued at $100–$200 million annually, with costs that were passed on to EU consumers. The association implied that outside pressure (by the United States and Latin America) would bring about a more substantive change in the EU policy (Agro Press 1998, 309–12).

Panama also was unhappy with the new regime. The proposed quotas for Latin America under Council Regulation 1637/98 would give the country only 15.76 percent of the first two quotas, much less than its 20 percent pre-quota share. Panama's sales to the lucrative EU market had declined by 40 percent since 1993. While the country had a diverse, service-oriented economy, its most important export was bananas. It did increase banana exports to the United States during the 1990s, but it did not realize the same prices from that market as it did from the EU (Newsome and Wilson 1999, 1).

The United States considered the new council regulation incompatible with WTO rules regarding tariff-rate quotas and the licensing system. On November 10, 1998, the USTR announced that the United States would seek retaliation options against the EU regime. The EU labeled this U.S. action as "unilateral" and a "serious departure from the WTO dispute settlement system" (Reuters 1998, 2), and it threatened retaliation within the WTO if the United States persisted on that path. The EU insisted that its new regime corrected those aspects of the previous system that were inconsistent with WTO regulations (Reuters 1998, 1) and defended its continued use of tariff-rate quotas rather than a tariff-only regime, noting that TRQs were commonly used in agricultural trade, including by the United States itself for sugar and dairy imports.

Council Regulation 1637/98 was submitted to the DSB, which ruled that the policy was indeed not fully WTO-compatible. The panel found fault with two particular aspects of the regime. It ruled that the use of the 1994–96 historical reference period for the issuing of licenses perpetuated market distortions present in the pre–1993 EU banana importation system. In addition, it decided that the separate ACP quota still violated WTO rules (European Commission 1999, 2–3). Frustrated with reviewing a still-incompatible policy after giving the EU fifteen months to develop a new one, the WTO in December 1998 authorized the United States and Ecuador to prepare a list of sanctions to be levied against the EU. As the level of confrontation escalated at then end of 1998, the North American press and public began to pay attention to the banana war for the first time.

The United States Levies Sanctions

After the WTO ruling on the revised EU banana regime, the United States announced a broad list of possible imported products that it would subject to sanctions when final WTO authorization was received. The list included items having no relationship to bananas and was valued at $520 million, well above the sum of $191 million initially authorized by the WTO. European businesses with goods on the list faced uncertainty about their future ability to export to the United States. Some businesses, such as Shearer Candles of Glasgow (UK), began to lose income because their U.S. buyers reacted to the threat of sanctions and directed their orders elsewhere; others, such as France's Societe des Caves de Roquefort and Louis Vuitton, reacted with "incomprehension" because their products did not compete with U.S. manufacturers or had no connection with the banana trade (Reuters 1999, 1). Such reactions were just what the United States desired when it announced its long list; its hope was that those

European businesses would work with other influential concerns, particularly importing firms that handled third-country bananas, to oppose the EU regime. The official EU response was offered by UK Trade Minister Brian Wilson, who stated: "There is absolutely no connection between bananas and cashmere or greeting cards except in the minds of the U.S. officials who have drawn up this list for random retaliation" (BBC News 1999a, 2).

Anticipating final authorization of its right to levy sanctions, the USTR in April 1999 announced the final list of products to be subjected to a 100 percent tariff. Several items gaining attention on the original list—notably Scottish cashmere and French cheese—were not included, an indication of the chain rattling that the United States hoped to accomplish by publishing the longer list in December 1998. However, the new list included such items as bath preparations; handbags; uncoated felt paper and paperboard in rolls or sheets; folding cartons, boxes, and cases of noncorrugated paper or paperboard; lithographs on paper or paperboard, not more than 0.51 millimeters in thickness, printed not more than twenty years before the time of importation; bed linen, not knot or crochet, printed, of cotton, not containing any embroidery, lace, braid, edging, trimming, piping, or appliqué work; lead-acid storage batteries; and electrothermal coffee or tea makers, for domestic purposes, except those from Italy (USTR 1999a, 1–2).

On April 12 the WTO banana panel published its final report on Council Regulation 1637/98, concluding that the policy was still inconsistent with WTO rules and authorizing sanctions. It gave the United States ten days to submit a list of goods that did not exceed $191 million in value. Peter Scher, the chief U.S. negotiator for agricultural trade affairs, stated that the list would reflect the U.S. goal of exerting "maximum political pressure" on EU politicians to bring about a change in the banana regime (BBC News 1999b, 2). On April 19, 1999, U.S. WTO ambassador Rita Derrick Hayes officially requested WTO authorization to suspend concessions (i.e., raise tariffs) on the USTR's final list, noting that "this is the first time this has occurred in the WTO." She went on to state:

The suspension of concessions . . . is essential to a major objective of the Dispute Settlement Understanding (DSU): prompt compliance with the rules of the WTO agreement. The arbitrators have recognized this fact and agreed that it is the purpose of countermeasures to induce compliance. After many rounds of litigation and exhaustion of almost all the procedural possibilities of the DSU, we have arrived at a stopping point for litigation on this issue. It is time to draw a line. If the rules of the WTO agreement are to be meaningful, there must be consequences for a failure to come into compliance with those rules. The DSU provides those consequences, provides legal safeguards with which we have complied in full, and guarantees that action will be taken. (Hayes 1999, 1–2)

The DSB responded rapidly, approving U.S. plans to suspend concessions on the products listed. The tariffs could remain in effect until such time as the dispute was resolved or the EU demonstrated that its banana regime was WTO-compliant. Charlene Barshefsky, then USTR, summarized the U.S. position: "This action validates what the United States has been saying . . . the EU has not complied with its obligations and this failure is damaging to the United States. Our action today redresses the longstanding imbalance in WTO rights and obligations and sends a clear message to the EU that *protectionism has a price*. We urge the EU to comply fully with the many WTO rulings against it on this issue" (USTR 1999b, 1).

U.S. sanctions went into effect in late April 1999. Ecuador also gained authorization to levy sanctions but preferred not to do so, choosing instead to pursue negotiations to accomplish its desired changes in the EU banana regime. Ecuador often stood alone during the banana war; it continued to do so in 1999 and thereafter. It opposed any quotas for ACP or Latin American producers and argued that its exporters lost two hundred million dollars annually to a licensing system that forced them to purchase licenses from EU firms. Ecuador welcomed the April WTO ruling on sanctions but was skeptical about its benefits; it had won the 1997 case only to see the EU spend fifteen months developing another wTO-incompatible regime. In 1998, 29 percent of Ecuador's banana export earnings came from sales to the EU, a figure that would rise substantially without the constraints of quotas and license systems (Newsome and Wilson 1999, 1).

The imposition of sanctions exacerbated divisions within the EU, as the United States had hoped that it would, with patience running thin. Many wondered if Brussels was incapable of creating a banana regime compatible with wTO rules. But the sanctions also generated another flurry of activity toward the goal of designing a regime that would bring an end to the war.

The First-Come, First-Served Controversy

By mid-1999, as U.S. sanctions took effect, the complexities presented by the banana dispute must have seemed insoluble to the European Commission functionaries who were responsible for bringing an end to the crisis. There were major divisions over each aspect of the issue, both within and outside of the EU. Within the EU, the gap between countries favoring free trade in bananas and those trying to preserve ACP preferences remained wide. Outside the EU, a new gap emerged among ACP states over the global ACP quota issue. Caribbean ACP countries opposed tariffication, feeling that it would not be possible to negotiate a sufficiently high tariff to protect Caribbean production (CBEA 1999, 3). The EU initially agreed, arguing that tariffs alone, unless they were very high, would be insufficient to protect EU and ACP banana producers from Latin American competition (*Journal of Commerce* 1999, 1). Among third countries, wide divisions remained between the BFA group and other exporters; Ecuador and Panama both pursued individual agendas on the dispute. How could the EU develop one wTO-compatible policy to satisfy such different priorities? It appeared that no one policy could accomplish the seemingly impossible task.

In November 1999 the European Commission proposed yet another banana regime, this one reflecting some well-hidden wisdom.

Unable to satisfy everyone, the EU presented a policy that was virtually guaranteed to satisfy no one. In retrospect, it appears that it did so as a means of getting everyone back into the negotiation process. In the interim, however, it generated quite a storm.

The proposed policy, known as the "first come, first served" (FCFS) system, would have two stages. Stage 1 would continue the quota system already in place, but now it would have a tariff of seventy-five euros per ton for 2.553 million tons of third-country bananas in the first two quotas and duty-free entry for an ACP quota of 850,000 tons with no country quotas. The EU, however, would seek agreement of all parties involved to "avoid further WTO challenges" (European Commission 1999, 1). Stage 2 of the proposal cited January 1, 2006, as the deadline by which TRQs would be dismantled, replaced by a tariff-only regime, with the level of the tariff to be determined at a later date. The controversial aspect of the new regime involved its license distribution system, which, in the absence of an agreement among all parties involved, would "be an appropriate form of a First Come, First Served (FCFS) system. The practical application of the FCFS system could present a number of technical and administrative difficulties, which would need to be overcome in such a way that the scheme is both physically practical and demonstrably non-discriminatory" (European Commission 1999, 1).

An FCFS system relies heavily on short-term quota management. The European Commission proposed either a fortnightly or a weekly basis for managing quotas to ensure a regular flow of bananas into the European Community. It would require operators to commit bananas to a vessel before submitting declaration forms of intent to import, a requirement intended to deter speculation. This step would be followed by a pre-allocation procedure, which would occur when the vessel was "a sailing distance from Europe" to avoid discrimination against countries that were farther away. All three quotas, including the ACP quota, would be managed this way, although ACP bananas would have a tariff preference of three hundred euros per ton for their 850,000-ton quota (Secretaria Técnica

Conjunta 2000a, 3). The proposal was very controversial, and the European Commission obviously knew that, but the commission expressed its frustration with the continual need to revise the banana regime. It stated that if no agreement had been reached by the time the European Parliament offered its opinion on the new FCFS policy, it would not *"be able to maintain the proposal* [emphasis mine] for a transitional regime. In that case, the Council should adopt negotiating directives to mandate the Commission to immediately initiate negotiations . . . with a view to replacing the current regime with a flat tariff" (European Commission 1999, 2). The European Commission showed its frustration to the Council of Ministers and the European Parliament in an effort to bend them to its will. A flat tariff was clearly unacceptable to the ACP states, and the Commission had long supported import regimes that protected ACP banana interests. The EU Parliament was not helpful, referring the issue back to the Commission and passing a resolution rejecting FCFS in April 2000 (CBEA 2000, 5).

With no guidance from the Parliament, the European Commission continued along the FCFS path. Its consultations with the parties involved in the dispute arrived at no solutions. In November 2000 the EU Commissioners for Trade and for Agriculture, Fisheries, and Rural Development submitted a final version of the FCFS regime to the Council of Ministers for adoption as a council regulation. They considered it a viable alternative to systems based on historic reference periods, on which no agreement could be reached, indicating that FCFS would bring an end to a dispute that "has already gone [on] for too long" and facilitate termination of U.S. sanctions (Secretaria Técnica Conjunta 2000a, 1–2).

The responses to the FCFS proposal were mostly negative. Caribbean ACP states vehemently opposed it, with the CBEA labeling it "a boat race" intended to avoid another WTO challenge (CBEA 2000, 3). In an effort to derail the FCFS policy, representatives of Caribbean ACP states met with the USTR. Agreement was reached on several points, including the two-tiered tariff-rate quota system and the

1995–97 reference period for granting licenses for the second quota. The licenses offered the necessary protection of access to EU markets, and U.S. acceptance of the license system and reference period facilitated agreement on an ACP tariff preference of 115 euros per ton, (CBEA 2000, 4).

FCFS would likely have different impacts within Latin America. Higher wage exporters like Colombia, Costa Rica, and Panama could be negatively affected, while Ecuador, with the lowest wages in the region, might benefit (US/LEAP 2001, 1). In the end, all Latin American exporters rejected the proposal—except Ecuador, which informed the EU that it could accept FCFS as long as it led to a tariff-only regime as quickly as possible. It also preferred that licenses under FCFS be allocated based on the "simultaneous requisitions of import authorizations." Banana workers' unions and many NGOs involved in the industry opposed the plan (US/LEAP 2001, 6).

The United States rejected the proposal one day after it was announced, on the grounds that the procedures for administering the FCFS system would favor EU firms. Not all reactions in the United States were negative. While Chiquita opposed the proposed regime, Dole viewed it more positively and said that it was willing to accept it as a transitional measure leading to tariffication (Secretaria Técnica Conjunta 2000b, 1–4). Dole sourced extensively in Ecuador and was better positioned to take advantage of the FCFS system.

The Council of Ministers supported the European Commission's proposal, approving Council Regulation 216/2001 on January 29, 2001, for implementation on April 1, 2001. Its Article 18 established three tariff-rate quotas, with a seventy-five euros per ton ACP preference in TRQs A and B and a three hundred euros per ton ACP preference in TRQ C (European Council 2001, article 18). Licensing procedures for the FCFS system were addressed with such oblique, nonspecific language that one must wonder whether the regulation's authors ever intended the system to be implemented: "To the extent necessary, importation of bananas into the Community shall be subject to submission of an import license, to be issued by Member States to any interested parties irrespective of their place of establishment within the Community. . . . Such import licenses shall be valid throughout the Community. . . . The issue of licenses shall be subject to lodging of a security against commitment to import on the terms of this Regulation during the period of the license's validity" (European Council 2001, article 17).

The European Commission might not have intended to implement the FCFS system. When presenting the policy, it also announced its "readiness to listen to proposals for an acceptable solution for all players involved" (EC 2001c, 2). FCFS was presented as an extreme option with potential negative impacts that would make other policy alternatives appear more reasonable by comparison. The policy was designed to provoke hostile reactions, which it did from nearly all parties involved in the dispute. That would draw them back to the negotiating table in a more compromising mood. The strategy seemed to work. Talks resumed in February 2001. They progressed rapidly to the point where many began to hope that the final battle of the banana war was being fought.

FOUR

Globalization

10

Pursuit of an Elusive Goal

The foreign debt crisis that affected LDCs during the 1980s stimulated globalization processes in the primary sectors of many Latin American and Caribbean countries. The need to service debts and qualify for new loans created an export imperative that led such countries to expand their commodities sectors as a means of generating foreign exchange. This dynamic coincided with greater application of free-trade principles to agriculture and the imposition of structural adjustment policies by foreign aid agencies and international lending institutions. As agriculture was forced to expand onto more marginal lands, the results were often harmful to local food production and to the environment.

In response to the export imperative, many countries looked for ways to diversify their economies. Some developed nontraditional agricultural exports (NTAES); others expanded production in traditional export sectors. Thrupp (1995) analyzed the growth of NTAES in Latin America, linking it to technological change and the penetration of TNCs into LDC agricultural sectors. The NTAE strategy seemed like a sound way to reduce dependence on the narrow range of traditional exports that were often subjected to fluctuating market prices, but in the long run the strategy was fraught with uncertainty. According to Thrupp, as more indebted states entered the fray, the competition among them increased, and the situation that already existed in traditional export sectors was replicated. Furthermore, the trend was an example of how LDC economies were once again being structured to serve the needs of the North. Adding to the risk of this strategy was the fact that one country's NTAE might be another's major traditional crop, a potential problem where bananas were involved. ACP countries with uncompetitive banana industries could face competition from large-scale producers like Brazil and South Africa, which previously had not marketed bananas in Europe.

The need to be competitive was perhaps never greater. As protective policies in agriculture were dismantled and trade preferences expired, pressure mounted on noncorporate farmers in the LDCs to increase productivity, efficiency, and quality. Since the late 1980s, banana farmers in ACP states had been subjected to continual EU criticism of the poor quality and high cost of their fruit. The origins of this criticism date back to Hurricane David (1979), when cheaper Latin American bananas replaced lost eastern Caribbean exports in the UK and French markets. The cause of the criticism about quality is not always clear. Many European consumers prefer the smaller bananas coming from the Caribbean. Although their appearance may not measure up to Latin American standards, they are sweeter and their taste is often viewed favorably. Nevertheless, the BGAS conveyed EU criticisms of Caribbean bananas to their farmers, who often interpreted them as the cause of the higher quality standards they were forced to fulfill. The imposition of those standards transferred more of the post-harvest handling process of bananas to the farmers themselves, adding to their workload while increasing their costs and providing less remuneration.

Welch (1996, 329) questions the relationship between quality and efficiency with regard to eastern Caribbean banana production. She suggests that increased efficiency was achieved during the 1990s, when smaller, less productive farmers were driven out of the industry. The cost of sustaining so many small producers was real; she notes that St. Lucia's thirteen inland buying stations and six input stores primarily benefited the smaller producers, who also held about 60 percent of the SLBGA's outstanding debt. But would the elimination of those producers yield quality improvements? There is nothing to suggest that it would, and BGA officials in Dominica, Grenada, and St. Lucia uniformly lamented the decline in the number of banana farmers that occurred during the 1990s.

Elements of Competitiveness

What is competitiveness and why is it increasingly valued in the world economy? Theoretically, competitiveness is the efficient use of resources, both human and natural, in ways that should yield less expensive products, making them accessible to greater percentages of the population. The concept is related to comparative advantage theory, the classical economic trade theory suggesting that everyone benefits when each country or region focuses on production of those goods for which its resources are best suited, leading to efficient production in terms of cost. Efficient production contributes to increased trade, with countries selling the goods that they produce most efficiently and importing goods that their resources are not well suited to produce.

Unfortunately, most research on competitiveness focuses on secondary and tertiary activities, not on agriculture. The *World Competitiveness Yearbook*, published by the International Institute for Management Development (IMD), reflects this bias. Its annual competitiveness survey ranks primarily developed countries, especially OECD states (IMD 2000). In 2003, however, the rankings were expanded to include several larger LDCs and eight subnational regions in larger nations. The 2003 yearbook for the first time acknowledged the significance of size, separating countries into two groups—those above and those below a population of twenty million (IMD 2003). The only banana-exporting states in the rankings were Colombia, Mexico, and Venezuela.

Stéphane Garelli, director of the World Competitiveness Project, identified several factors contributing to a country's competitiveness level, including socioeconomic environment and the ability to attract investment capital (Garelli 2006, 10–11). He notes a distinction between an "economy of proximity" and an "economy of globality" (Garelli 2006, 12). The former is composed of traditional activities, government, and personal services in which value is added close to the end user of the good or service. The latter is dominated by international firms for whom proximity between production and consumption is not important but competitiveness and efficiency are. When applied to the banana trade situation, Garelli's ideas are far more relevant to the Latin American industry than to the ACP producers. The latter are less able to attract capital, and their stock is declining within a milieu of liberalizing trade regulations. Also, the ACP banana industries do not represent true economies of globality because they do not control those aspects of the industry that lie beyond their national frontiers. Garelli offers several "Golden Rules of Competitiveness" that are applicable to diversification, including the creation of a stable legislative environment, investment in both traditional and technological infrastructures, transparency and efficiency in government, strengthening of the middle class and preservation of social cohesion through reduction in wage disparity, and investment in education and job training (Garelli 2006, 15–16). Institutions such as the IMF and the World Bank, which ostensibly encourage competitiveness in the developing world, actively negate several of the "golden rules" through their structural adjustment programs. This is particularly true where public expenditures are involved, as they are in education, infrastructure development, and strengthening of the middle class.

One important element of competitiveness, regardless of sector, is the ability to generate economies of scale. Its underlying principle suggests that greater production volumes lower the unit cost of production, thereby reducing possible sale prices. This dynamic occurs because fixed costs can be spread over a larger volume of production, contributing to increased efficiency. Although other aspects of production might be very efficient, the inability to amortize fixed costs over a larger volume will always present a disadvantage to small-scale producers. Market size often affects the economies of scale picture. A small domestic market, which results from either a small population or an unequal distribution of wealth, is a disadvantage, but that disadvantage can be offset by production for export to larger markets abroad. The Swedish example of Volvo production illustrates this concept: the small internal Swedish market is partially offset by international demand for a quality product, enabling the company to produce and sell more Volvos at a more competitive cost. Thus, access to foreign markets is a critical issue for countries with small domestic markets, rendering the loss of Lomé Convention preferences an even greater blow to the ACP states.

Another important aspect of economies of scale is the internal structure of the production system. Where production is structured around a large number of small-scale producers, costs are likely to be higher, within the same social context, than in situations where larger-scale production units dominate. Thus, the family farm system of banana production in the eastern Caribbean works against the derivation of economies of scale, rendering them uncompetitive.

Competitiveness and Agriculture

How can competitiveness in agriculture be evaluated? Of the competitive elements noted above, economy of scale is particularly applicable in agriculture. As the scale of the production unit increases from family farm to plantation or corporate farm, mechanization and infrastructure levels usually increase. Those inputs cost more at first, but the ability to amortize over greater production volumes ultimately lowers per-unit costs and the price charged for the good produced. Small-scale farms also encounter higher costs in other areas, such as inputs, especially if they are imported due to insufficient demand volume to allow domestic production. Transportation is also affected, with lower volumes leading to higher transport costs. These issues lie beyond the control of the individual producer, although cooperation can offset some of the disadvantage of small size.

Other elements of agricultural competitiveness are not directly related to scale. Significant among these is the suitability of the physical environment to the particular endeavor. Factors such as soil fertility, climate, the presence of pests, and moisture availability affect the need for additional inputs (e.g., fertilizers, pesticides, and irrigation), either raising or lowering production costs. Topography, another physical factor, interacts with production scale; irregular topography favors smaller production units but makes it difficult to derive economies of scale.

Human factors enter into the picture, although these tend to vary more between countries than within them where similar social norms and rules apply. Expenditure for labor, both in the form of wages and in social assessments imposed by governments, represents an important element within the production cost structure. The need for training or retraining of workers, whether proletarian wage earners or family farm owners, adds to the cost of production.

With the end of splendid isolation, such issues of competitiveness are now more widely studied in many LDCs by state agencies and IGOs. A CARICOM study of regional agricultural systems identified several structural weaknesses and external constraints that were preventing the banana sector from fulfilling its potential (CARICOM 1996, 9–10). Topping the list was the numerical dominance of small farmers with limited technology at their disposal. The study noted that limited technology contributed to inefficiency and limited competitiveness. Other problems identified included poor procurement systems, many small traders handling small product volumes, small domestic market size, inadequate transportation, insufficient investment in research and development and in agricultural extension activities, inefficient produce inspection and quality certification, poor post-harvest handling systems, and inadequate manpower for international marketing management. In relation to the last item, the report also noted the absence of a regional entity charged with responsibility for joint extra-regional marketing of CARICOM agricultural exports.

These problems are felt by virtually all countries in the region. Their importance is clear in the following discussions of competitiveness and economic diversification.

Competitiveness and the Banana Trade

Competitiveness is a primary issue underlying the banana trade dispute. The various banana production systems yield different outcomes with regard to cost and efficiency. Latin American producers are more efficient—in terms of productivity and cost, if not in longer-term social and environmental outcomes—than ACP and EU producers. The difference in efficiency was less important while Europe maintained its segmented market in bananas with its Lomé Convention preferences for ACP exporters. However, the advent of the SEM and the creation of the WTO brought an end to the security offered by that system.

The EU tariff-rate quotas for third-country bananas contributed to competition within Latin America shortly after the implementation of Council Regulation 404/93, when Ecuador began selling its bananas at very low prices (Sánchez 1993, personal communication). The competition affected the earnings of higher-cost Latin American producers such as Costa Rica and Panama. Although EU documents customarily discuss Latin American producers in terms of one homogeneous dollar zone, there are significant differences among them, particularly with regard to labor costs. Colombia, Costa Rica, and Panama have the highest wage rates among major

TABLE 7. Labor costs and unionization rates for banana industries in
selected Latin American countries

Country	Wages (US$ per day)	Unionization rate (%)
Colombia	9–11	90.00
Costa Rica	7–10	6.38
Ecuador	1.25–2.50	0.05
Guatemala	4–5	41.40
Honduras	5–7	76.16
Panama	17–21	74.00

Source: US/LEAP 2001, p.5

Note: Solidarismo membership, especially high in Costa Rica, is not included in these totals.

Latin American exporters (table 7). Kieswetter-Alemán (2000, personal communication) attributes this to the fact that workers' salaries in the three countries' banana sectors are higher than salaries for other agricultural activities. In Panama an additional 28 percent of workers' salaries is paid in benefits, mostly to the government to cover the cost of social security, unemployment compensation, health care, and other social programs. In Costa Rica the figure is 24 percent. Ecuador is at the other extreme; salaries there are lower and workers receive fewer benefits. In addition, few workers are unionized.

Overall, the EU banana policy had a negative impact on labor in the Latin American exporting countries. Its tariff-rate quota for third countries contributed to an oversupply in global markets that drove banana prices down, causing earnings to slump for most sectors of the industry and leading to cost-cutting measures with serious repercussions for the workforce. The strength of unions waned; producers used subcontract labor rather than employed workers, further deflating salaries and benefits. Perillo (2000, 4) described the climate for organized labor as increasingly unfriendly, as governments attempted to make their economies more competitive at the expense of labor.

In Ecuador the unions' difficulties began with pressure on the government from the banana companies. Today the industry has a non-union workforce that ranks as the most poorly paid in the region. Cheap labor provides Ecuador with a natural competitive advantage in global markets and explains its independent behavior with regard to the trade dispute. Many consider Ecuador to be the winner in the "race to the bottom" of the banana world, a situation that cannot be blamed on U.S. TNCs. Ecuador's industry is dominated by the world's fourth-largest banana company, Noboa, which is owned by an Ecuadorian national.

Government subsidies also affect competitiveness among Latin American producers, with Ecuador again the exception. The government of Ecuador subsidizes the cost of Sigatoka negra control, which averages fifty cents per box of bananas. The subsidy makes it difficult for other countries to compete with Ecuador as the target cost to do so is lower than their production costs (Kieswetter-Alemán 2000, personal communication).

The Challenges of Diversification

The legacy of specialized economies carried over from the colonial pasts of most LDCs became the target of many diversification efforts throughout the developing world. Two of the motivations for this trend in Caribbean and Latin American countries were the onset of the foreign debt crisis and the EU banana importation policy. The former increased the necessity of boosting foreign exchange earnings, and the latter threatened the demise of a major export sector in several countries. Ironically, diversification is encouraged, if not mandated, by the very countries responsible for the colonial and neocolonial structures that created the highly specialized economies in the first place. A further irony is that diversification efforts increase the difficulty of achieving competitiveness for any one good.

Nevertheless, a diversified economy is clearly preferable to a specialized one. The latter is more vulnerable to external shocks and

19. The Dominica Export Agency's optimistic array of
diversification products in Dominica.

natural hazards. It is subject to price fluctuations and boom-and-bust
cycles characteristic of most commodity exports. A high degree of
specialization renders other sectors, particularly tertiary activities,
more dependent upon the well-being of the principal export commodity. This is illustrated at national and subnational levels in the
eastern Caribbean and the banana zones of Latin America, respectively. Diversification offsets this tendency, reducing the amplitude
of boom-and-bust cycles by decreasing the likelihood that all activities will experience such trends simultaneously and improving the
number of employment options available.

As a result, economic diversification is now a policy priority in
most LDCS. The trend toward NTAE development is a significant element within this process. Conceptually, diversification seems an obvious choice for national policymakers. However, in reality it is a far
more difficult goal to attain than may be immediately obvious. The

challenges inherent in the diversification process are discussed in the context of the eastern Caribbean ACP countries for three reasons. First, several ACP countries involved in the banana dispute remain considerably more specialized than most of their Latin American competitors. Thus, they have a more difficult road to travel to diversify. Second, bananas are more central to the social fabric of those countries. Third, the ACP states encounter the dual challenges of higher labor costs and smaller size, with the attendant issue of economies of scale, factors that substantially affect their diversification efforts. The Latin American banana exporters, on the other hand, are characterized by a higher degree of economic diversification at the national level. Although their banana zones remain quite specialized, they are rather peripheral to the consciousness of the national societies involved. They are also much larger than the Caribbean ACP states, have lower labor costs, and do not encounter the same degree of problems of economies of scale.

What is involved in diversification, once the decision has been made to pursue such a process? At a minimum, the process includes several dimensions, the first being the decision about which sectors to target in the effort and the second being the mix of products to be included in the diversification package. The important selection process raises questions about the appropriateness of the resources available, the infrastructure requirements of each new industry, labor and training needs, the inputs required and their source(s), the potential markets for the new goods and access to them, the transportation links that will be needed, who the competitors will be, and what kind of marketing strategy must be developed. These questions and many others must be answered for each product considered for inclusion in the diversification mix. Thus, diversification is a complex and slow process (Wiley 1998). It is often costly to implement and it is fraught with risks, including many of the same problems a country encounters with its traditional exports in an increasingly competitive, globalizing economy: price fluctuations, market access, natural hazards (in agriculture), and competition from more efficient producers. With diversification, however, these problems are encountered in less familiar environments than those associated with a country's traditional exports.

State officials, private-sector representatives, NGO workers, and farmers inevitably express a mix of opinions about diversification, with any optimism about outcomes tempered by the reality of obstacles to be overcome. In Dominica, Carey Harris, director of the Diversification Implementation Unit (DIU), suggested that, in Dominica, the government's role in the diversification process included filling gaps in education, training, and entrepreneurial skills, as well as overcoming the private sector's aversion to assuming risk. He indicated that diversification in Dominica would be gradual and that market stability would be achieved for some of the products in Dominica's mix but not others, especially not those with extra-regional markets (Harris 1995, personal communication).

Dr. Donovan Robinson, chief technical officer for Dominica's Ministry of Agriculture, added that the government of Dominica could provide incentives to convince farmers to diversify their crops. Needed incentives would include assurances to farmers that potential markets exist, the creation and subsidizing of plant stock for new crops, and provision of credit to the farmers engaged in the new endeavor (Robinson 1995, personal communication). He identified shipping as a great obstacle for Dominica because the country has few direct shipping links to major markets, so it is forced to rely on intermediate stops (e.g., in Antigua) for transshipment.

Convincing farmers to switch from bananas is not easy. Many farmers express skepticism about other crops, most of which are seasonal and do not provide the year-round income offered by bananas. They also fear an extended transition to the new crop(s), during which they would earn little or no income. They also must learn new techniques to cultivate, harvest, and pack the new crops. Many of the region's farmers are relatively old and do not welcome such changes to their means of earning a livelihood.

Strengthening links between agriculture and tourism so that the former can supply the latter is another role the state can play in diversification. Heather James, Dominica's chief technical officer in the Ministry of Finance, Trade, and Tourism discussed this need, noting that more must be done at the institutional level to help this process along (James 1995, personal communication). She also noted that tourism can be a test market for a range of agricultural goods in a diversification mix, such as fresh fruits and vegetables or cut flowers, although it is a challenge to make producers aware of such opportunities.

Defining a list of diversification crops in the eastern Caribbean involves trying to determine which goods the region could produce competitively. Frank McDonald (1995, personal communication) noted that the Caribbean has "no monopoly on anything," so its choices would have to be judicious and flexible. He considered adoption of appropriate technologies to be a key to a diversification strategy in the Caribbean's agricultural sector, enabling it to better compete on world markets. He suggested that the islands look within traditional sectors for additions to their diversification export mix. This could involve agroprocessing, such as converting bananas into baby foods or chips. However, Tim Durand questioned whether sufficient markets could be found for the secondary sector goods resulting from such activities (Durand 1996, personal communication). The theme of diversification within traditional sectors has been echoed in Grenada and Belize. In Grenada nutmeg and cocoa are the most important traditional crops, and bananas are next. Nutmeg prices declined during the 1990s, reducing profits for many farmers. To counter this, Grenada established a nutmeg distillation plant on the island's northern coast in 1996. It produces nutmeg oil for use in cosmetics, preservatives, and seasonings. According to Anthony Boatswain of the Grenada Industrial Development Corporation (GIDC), the oil is a value-added good, but the eventual goal is to diversify into the even more profitable production of goods that use the oil as an input (Boatswain 1997, personal communication). In Belize, where legislation fostered the creation of export processing zones, dehydrated bananas for use in wheat flour and baby food have been under consideration (Martinez 1998, personal communication). Such exports, however, would lack the same protection as the bananas themselves (Flores 1998, personal communication). The similarity of these ideas across several countries suggests that, should the ideas be translated into action, new forms of competition would supplant the old, reducing any eventual benefits.

Conrad Cyrus of the Dominica BGA noted the time needed for the implementation of agricultural diversification, particularly with tree crops that require several years after planting before they begin to bear fruit. Where bananas provide a relatively quick return on investment—usually nine months—mangoes and citrus require three to four years and coconuts at least seven years. For small-scale farmers with limited incomes, the waiting period is a disincentive for change (Cyrus 1995, personal communication). The problem is exacerbated by the large number of small-scale farmers in the region, which slows down the conversion process and increases the difficulty of quality control in the new activity (Martin 1995, personal communication).

Perhaps the best indicator of the true challenge of diversification to several ACP states is that it has been discussed for so long while producing so few results. J. M. Marie (1979, 74–76) presented an early analysis of many of the issues, including the limitations of small market size. Thomson (1987) made diversification the centerpiece of the concluding chapter of his book on banana production in the eastern Caribbean. The truth of the matter is that whatever diversification efforts might be pursued, they always encounter many of the same problems that afflict the banana industry. Bananas represent the reality confronted by the ACP states. Banana farmers control nothing beyond the production stage, which bears the greatest risks, particularly from unpredictable weather. Foreign concerns control the shipping, the marketing, and the pricing structures driving the industry. Foreign governments—in this case, the EU—control access to markets through a preference system that came under attack by another government, the United States, representing the interests of competing TNCs. The dependency aspects of the banana trade dispute would carry over into virtually any other development path that the ACP states (or Latin American countries) might decide to pursue. It is unlikely that true competitiveness, as it is traditionally conceptualized, can be achieved through diversification by countries that control so few of the resources and make so few of the important decisions that would determine their ultimate success.

Under Pressure: The Small Economies

Diversification is complex and difficult for any society, but for very small countries the process is an even greater challenge. The special problems faced by small economies have gained increasing recognition in international venues like the United Nations, where the UN Conference on Trade and Development (UNCTAD) takes a leadership role, in EU-ACP cooperation mechanisms like the Cotonou Agreement, and even within the WTO, although it remains unclear whether its attention will translate into policies that actually address the needs of the countries involved. The International Institute for Management Development's inclusion of a small economies category reflects the increased recognition and was reinforced by Stéphane Garelli's comment that "the competitive approach must be managed differently in small economies than in larger ones" (Garelli 2003, personal communication). Garelli suggested that large and small economies attract different types of investments. Larger economies attract market-serving investment that flows to industries producing for domestic markets. Smaller economies, lacking significant markets at home, must attract market-sourcing investment oriented toward production for export.

Very small states face environmental challenges and limitations that affect their economic development. These include a limited natural resource base, extreme susceptibility to natural disasters, a high degree of dependence on international trade, high transportation costs due to low volumes, a limited tax base to pay for infrastructure, high per capita public administration costs, and most significantly, an inability to generate economies of scale (SIDSNet 2000, 1). The inability to generate economies of scale is a central problem. It affects performance in several areas related to international trade, particularly the difficulty small countries have in producing goods at competitive prices and transporting them to markets in a timely, efficient manner.

In the eastern Caribbean, political actions interacted with economic factors to contribute to the problem of economies of scale. During the late stages of the colonial period, when the islands were self-governing, many large estates were dismantled and land was distributed to the growing numbers of peasant farmers. The process continued after independence, giving more people a direct stake in the societies that emerged. It reflected the fundamental philosophy of empowerment advanced by leaders like Jamaica's Michael Manley and led to banana farming as a way of life for thousands of Caribbean farmers (Brown 2001, personal communication). But it also decreased the average farm size to just three acres. So while the system fostered democratic values and political stability, it also made economies of scale in agriculture more difficult to achieve. Small-scale firms also dominate the manufacturing sector in countries such as Grenada, where the largest plant employs just three hundred people. Along with higher than average wages and social payments, the difficulty of achieving economies of scale makes Caribbean ACP states less likely to become competitive for goods whose competitiveness is determined by production costs and price.

Transportation presents another major obstacle to competitiveness in small states. The issue has two primary dimensions: cost and availability. Small economies are characterized by high per unit costs to transport goods to overseas markets. Distance, foreign ownership of shipping lines, and the need for ships to call on several small countries to fill their cargo holds before heading to northern markets all contribute to higher costs. Each individual small island state is unable to generate sufficient volume to merit its own direct shipping link to markets in European or North America. The four-stop pattern represented by banana shipments from the eastern Caribbean illustrates this, as did Belize's need to export bananas via Honduras. It is often necessary to ship a combination of crops and other goods in order to make shipping economically viable. In recent decades, the rise of containerized shipping has worsened matters because shipping firms are reluctant to underuse containers.

In order to be successful, a transportation system must have several elements in place. The system must include packing of products to be shipped, movement of goods to port, storage facilities at the port, and the seagoing vessels to take the goods to market. Such a system must be available for each good included in a diversification product mix, and it must be appropriately timed, based on the perishability of the good (Robinson 1995, personal communication). While the first three elements can be controlled internally, the last is often problematic. Availability of suitable international trade links is often bemoaned as a critical problem facing small countries (Satney 1995, personal communication). The problem extends to air transportation as well. St. Vincent and Dominica lack airports capable of receiving large jets, necessitating an intermediate stop, often in Barbados or Antigua, for airfreight going to North America or Europe. Unless refrigerated storage is available, the extra stop can create spoilage problems for agricultural goods if delays occur at the external airport.

One example from Dominica illustrates the shipping availability problem and its impact on competitiveness. Bello Ltd. is an agroprocessing firm that produces for export. Although most of its markets are in developed countries, it receives some orders from neighboring islands. Vantil Fagin, company proprietor, explained his problem in attempting to ship goods to Guadeloupe, which is visible from the country's northern coast. The ferry linking the two islands carries passengers only (Fagin 1996, personal communication). Fagin's best option was a North American company that would carry his cargo to Guadeloupe via Miami. According to Fagin, inter-island shipping availability was "dreadful."

The transportation system for bananas, however, is already in place and could be used to piggyback the transport of diversification goods. A system to accomplish this was taking shape by 1995 in Dominica, where the Dominica Export Import Agency (DEXIA) was trying to develop markets for the country's new exports. However, the reality is that these goods must go where the bananas go—in this case the UK—or be transshipped from the UK to subsequent markets on the European continent (Clarendon 1995, personal communication). CARICOM's legal advisor to the WTO captured the essence of this situation, citing the words of John Compton, former prime minister of St. Lucia: "When people talk about diversification, they fail to realize the integrated structure of the economy. You need ships to export. Without bananas, there are no ships" (Brown 2001, personal communication).

Although this analysis has focused on the negative aspects of small economies, there are advantages to small size. The requirements for success reflect the scale of the economy. With a small population base, fewer jobs, in absolute numbers, are needed than are needed in Mexico or South Africa for a diversification program to be a success. Similarly, with a small volume of goods to ship, a country like Dominica can establish an arrangement with one large retail chain or packaging house in a market country to accomplish all of its export needs (Harris 1995, personal communication). Despite these advantages, small size seems to work mostly against achieving competitiveness, diversification, and survival in a neoliberal global economy.

The Niche Market Strategy

Although there appears to be no perfect solution to the problems that small economies encounter, the niche market strategy seems to be the approach best suited to confronting the challenges. The strategy is widely pursued by small countries today. There are three categories of niche markets that apply to the situation at hand, including temporal niches during down times in the export cycles of major suppliers, ethnic niche markets created by diaspora migrations, and markets for luxury or specialty goods in which qualitative factors rather than price determine the ultimate demand for the good in question. Temporal niches often involve seasonal agricultural harvest cycles, although they can apply to a country's tourism product as well. In agriculture, Dominica can fill the gap left each autumn by the downtime in Israel's citrus exports to the EU. In tourism, winter holidays in the Caribbean can attract European and North American sun seekers.

Ethnic niche markets have grown significantly since immigration laws changed in Canada (1962) and the United States (1965). Coupled with decolonization, the change in the laws has enabled immigrant groups from many Caribbean states to achieve the critical mass necessary to sustain ethnic restaurants, grocery stores, and shops in the destination countries of North America and Europe. Expatriate communities today represent important potential market niches for exports from their source countries. The issue remains, however, that any tropical good desired by an expatriate Caribbean communities can be and usually is produced more cheaply by a larger tropical country, or even Florida, making it necessary to compete on a basis other than price (Satney 1995, personal communication).

Many developed countries manufacture goods for luxury or specialty markets. Belgian chocolates, Swiss army knives, Milan fashions, and French perfumes all exemplify a strategy pursued under less dire circumstances than those now facing the world's small economies. In agriculture, organic farming constitutes a quality-based niche market approach. The specialty market strategy is now employed by banana growers in Costa Rica and elsewhere, yielding a quality product that does not directly compete with mass market fruit.

Niche markets tend not to be large markets, and this may be

the key to the strategy's success. They are often too insignificant for larger producers—that is, the competitive producers—to worry about. Without the larger producers, uncompetitive producers have space to maneuver and thus might have their best opportunity for survival during this era of globalization.

Carey Harris endorsed the niche market strategy in his diversification plans for Dominica, noting that the strategy offered the country's best chance to compete, as long as the country could maintain high quality and produce a consistent supply of the goods in its product mix (Harris 1995, personal communication). In the mid-1990s, Dominica took a scattergun approach to diversification, leaving no stone unturned as it sought niches in which it could succeed. In Grenada, Lucia Livingston-Andall and Gregory Renwick also viewed niche markets as the way to go; they believed that the development of exports considered exotic by upscale northern markets would provide a path to success (Livingston-Andall and Renwick 1997, personal communication).

Each potential niche carries with it all of the tasks and challenges inherent in diversification. Most niche markets for the small Caribbean states are necessarily extra-regional, which creates a transportation issue as well. But there are opportunities within the region that can be exploited. The Caribbean tourism sector represents a potential market within agriculture. Tourists are major consumers of imported items for which regionally produced substitutes could be developed. Developing this niche would necessitate strengthening the links between the agriculture and tourism sectors, which most parties agree is desirable. Within tourism, cruise ships are a potential growth market for Caribbean goods, including fresh fish, fruits, vegetables, flowers, and fresh water (Clarendon 1995, personal communication). Most of these goods are now supplied from North America by containerized ships that call on the homeports of the cruise lines.

Tourism itself offers potential for creating niche markets. The Caribbean is known as a resort-based tourist destination, but opportunities exist to develop niches within the industry. Dominica markets itself as the nature island of the Caribbean and wants to increase its profile as an ecotourism destination (Gregoire and Fadelle 1996, personal communication). Similarly, Belize is known for ecotourism and scuba diving; it desires to attract visitors interested in culture by participating in a regional *Mundo Maya* (Mayan World) joint itinerary development program with its neighbors (Woods 1998, personal communication). Many small states have name recognition issues that must be overcome through more effective promotion—an issue for Belize, Suriname, and Dominica, which is commonly confused with the Dominican Republic. Transportation is also problematic in countries in which nonstop flights from Europe or North America are few in number or impossible because the country lacks a jetport.

The niche market concept fits in nicely with the small scale of eastern Caribbean economies. The relatively limited number of jobs that are likely to be created will have a greater impact there than in larger countries, which must generate far greater numbers of new positions to achieve success. Thus, a niche approach can contribute to national development. Despite its appropriateness, however, the niche market strategy still faces problems that could undermine its long-term success. The obstacles include time, market access, market fragility, and, as always, transportation.

Time is critical to small states needing economic restructuring or adjustment. The niche market strategy presents the same obstacles as other diversification strategies. Small states argue in their multilateral negotiations that a phase-in time must be provided and that traditional exports must continue to have preferential market access until adjustments can be accomplished. They reiterate their support for trade liberalization but argue that the needs of small countries must be recognized. Thus, it is fair trade, rather than free trade, that is needed.[1]

Market access is another issue. There is no guarantee of access for new goods, so market development is extremely important. It is a costly activity in its early stages and must now be done in a milieu in which the rules of trade have changed. The preferences traditionally granted to bananas, sugar, and other goods will not be forthcoming. Niche market demand is often fragile, particularly for luxury or specialty goods for which demand is based on desire, not necessity. Tastes change frequently, leading to long-term instability and the need for continual revision of a country's export product mix. This instability can yield a potentially endless diversification process. Finally, the inherent transportation problems that plagued traditional exports for decades apply to niche market products and any other diversification exports that might emerge.

Outcomes of Diversification: A Race to the Bottom?

Where will diversification lead? The neoliberal world economy that grew out of the foreign debt crisis of the 1980s is often described as being characterized by a "race to the bottom" among LDCs. Critics suggest that neoliberal economic policies fostered the transfer of secondary sector activity to the LDCs by transnational companies seeking cheaper labor and weaker environmental regulation. The need to service debts and boost exports led many LDCs to compete with one another by providing incentives like tax holidays to lure foreign companies while simultaneously reducing public-sector expenditures on social services. *Maquiladora* districts and export-processing zones were created, offering favorable conditions to TNCs, including duty-free imports of the inputs used in the manufacture and assembly of finished products for sale abroad. At the same time, real wages often fell, labor rights were eroded, price supports for fuel and food were eliminated, social support systems were reduced or dismantled, and foreign control of LDC economies was reasserted. On the positive side, many women were employed outside of the home for the first time, and material consumption increased in many countries (although this can be attributed to a variety of factors). On balance, however, the situation seems to be a negative one in which

LDCs tried to outdo one another in their efforts to create conditions favorable to global capital at the expense of the local population. These conditions frequently led to riots, increased crime, and other manifestations of social instability.

The changing banana trade situation can be viewed as one more example of this dynamic, although this view oversimplifies the matter. From the ACP perspective, the U.S. challenge within the WTO favors its own fruit TNCs, which operate under advantageous conditions on the Latin American mainland. There, according to the EU and ACP perspective, a cheap workforce, working under difficult circumstances, produces bananas for export on land that they do not own. Of course, this view ignores the differences in conditions among the mainland Latin American producers. However, it does illustrate the significant difference between land tenure patterns in Latin America and those in ACP Caribbean states, and it highlights the generally better social conditions among the latter.

The potential for the traditional ACP banana exporters to join the race to the bottom clearly exists. The final outcome of the banana trade dispute ultimately will work to their disadvantage; the social fabric of those societies could be irreparably harmed. The late Eugenia Charles, former prime minister of Dominica, considered the banana trade dispute to be a matter affecting the survival of the country (European Commission 1993a). An inability to export bananas to the EU would lead to unemployment and consumption declines affecting all sectors of the economy. Even tourism would feel the impact of a less idyllic social climate. The region's governments would be forced to yield to the demands of international capital to replicate the conditions of competitiveness that now exist elsewhere.

Oliver Benoit, director of diversification for Grenada's Ministry of Agriculture, viewed the competitiveness imperative as the focal point of the problems facing the ACP banana-exporting countries. In his view, the global economic powers do not recognize that small countries can only achieve competitiveness through great sacrifice in the social arena. Achieving competitiveness would come at the expense of the independent farmers who are the bedrock of Caribbean societies. The likely outcome would be the loss of their independence and possibly their land. Should that result, he asked, what have we really gained by achieving competitiveness (Benoit 1997, personal communication)?

Dominica's articulated goals for its diversification program include, in order, maintaining democracy, increasing the standard of living of the country's farmers, and earning foreign exchange. DIU director Harris saw these goals as fostering socioeconomic development (Harris 1995, personal communication). Despite the difficulties of diversification in such a small country, Dominica hopes to maintain the social stability that has largely been lost in other developing societies.

However, as WTO involvement in the banana trade has grown, the prospects for such an outcome for Dominica and other ACP states have declined. With the loss of EU trade preferences and the erosion of income from their major export commodity, Caribbean ACP states now face economic shocks that will make it difficult to uphold their social democratic ideals. Should that occur, the citizenry of the region might become an example of sacrifice upon the altar of competitiveness.

11

Implications for the Future

By 2001 the banana war participants appeared to manifest symptoms of "banana fatigue." That fatigue, combined with the EU's threat to implement an FCFS policy that was clearly unacceptable to nearly all parties involved, motivated everyone to return to the negotiating table, and events progressed rapidly by the early spring.

April 2001: An EU-U.S. Agreement at Last

The approval of Council Regulation 216/2001 pleased no one. The EU again began a round of consultations with the parties to the banana dispute; although there was uncertainty as to whether or not the policy was WTO-compatible, the possibility that it might be compatible interjected an air of seriousness into the negotiations. In addition, the banana industry itself was suffering.

Banana prices declined during the latter half of the 1990s, reducing earnings for many companies involved in the banana trade. Stock prices for the U.S.-based TNCs plummeted during 1998 and

1999. Chiquita, Del Monte, and Dole declined by 60 percent, 71 percent, and 52 percent, respectively, all during a period when the overall Standard & Poor's 500 index rose by 16 percent (Perillo 2000, 5). These losses had negative consequences for labor. In 1999 Del Monte fired its entire workforce in Costa Rica, later rehiring most of the workers at reduced pay and benefits. In the same year, it fired nine hundred unionized employees from three Guatemala plantations and subsequently leased those plantations to Guatemalan nationals who hired nonunion workers at lower pay. Dole suspended operations in Nicaragua and Venezuela, terminating its contracts with independent producers there. Chiquita converted more Honduran plantations to African palm, which requires far less labor; it had employed this strategy previously in Costa Rica. In the United States, Chiquita announced that it was filing for Chapter 11 bankruptcy. All of these actions added to a growing sense of urgency regarding settling the trade dispute. Negotiations progressed rapidly, and in March 2001 the EU announced that it would not implement the FCFS policy, offering hope that a more acceptable policy would be forged.

On April 11, 2001, the United States and the EU announced that they had arrived at an agreement to end the dispute. It was a compromise, an accord covering a transitional period up to January 2006. By then the EU was to have completed the negotiations necessary for the transition to a WTO-compatible tariff-only regime (USTR 2001a, 2). The so-called tariffication system was acceptable to all parties, including the ACP states, as long as a sufficiently high tariff preference was established. In the interim, the EU would maintain the three earlier TRQs (Council Regulation 216/2001) but agreed to transfer 100,000 tons from TRQ C (the ACP quota) to TRQ B to be more accessible to third countries. In return for this concession, the United States agreed that the ACP states would have exclusive access to the remaining 750,000 tons in TRQ C (Lamy 2001, 2), and it also agreed to support EU efforts to obtain WTO authorization for the new accord (EC 2001d, 2), including the WTO waiver to reserve TRQ C for the ACP. Finally, the United States yielded on the use of a historic reference period as the basis for issuing licenses during the transition to tariffication. This concession was made in return for the EU agreeing to a 1994–96 reference period that would prove less damaging than a later period to the U.S.-based TNCS (USTR 2001b, 2).

In presenting the agreement, EU trade commissioner Pascal Lamy noted that the United States would suspend its sanctions on July 1, 2001, when the policy took effect. He summarized the EU perspective on the matter: "After many years of disputes, I believe that the aim which the Council and European Parliament set the negotiators have now been achieved: to protect Community producers and produce from the ACP countries while retaining compatibility with the EU's WTO commitments" (Lamy 2001, 2). The details of the new EU banana import regime were worked out during the remainder of April and were influenced by the simultaneous negotiations with Ecuador. The final policy was announced on May 2 and approved as Commission Regulation (EC) 896/2001 on May 7. It went into effect on July 1, 2001. Its most important provisions dealt with import license allocation methods, including specific definitions of traditional operators for handlers of third-country nontraditional ACP and traditional ACP bananas (Article 3). The provisions identified opportunities for nontraditional operators to qualify for import licenses (Article 6) and laid out detailed rules for the granting of licenses to traditional and nontraditional operators (Articles 13–21).

The agreement eliminated all individual country quotas, at least for the time being. It left open the possibility of establishing individual quotas within the first two tariff-quotas, if all of the major parties agreed to the quantities allocated to each country. That was never possible during the banana war but might still be feasible if Ecuador and Panama both signed on to the accord. The ACP quota also lacked country allocations.

Reactions to the Agreement

Reactions to the EU's new banana regime were varied; the array of interests to be served precluded unanimous approval on such complex issues. The reactions of the United States and Ecuador were the most crucial because they were the parties most responsible for the banana war's longevity. The USTR's favorable response was determined by Chiquita's positive reaction. The lack of country quotas in the new banana regime meant that licenses could be granted directly to companies rather than countries, and this had been a primary U.S. goal during the dispute. By selecting the earlier reference period, 1994–96, the EU ensured more favorable conditions for Chiquita, which lost nearly half of its EU market during the latter half of the 1990s and was on the verge of bankruptcy in early 2001 (US/LEAP 2001, 6). For these reasons, the company welcomed the accord. Since it was widely considered to have been the driving force behind the U.S. banana policy, Chiquita's favorable response to the policy removed the U.S. government as an obstacle to gaining acceptance for the accord.

Dole, on the other hand, had learned to play the 404/93 system more effectively than Chiquita had. In addition, Dole sourced extensively in Ecuador, so it was better positioned than Chiquita to benefit from the FCFS policy. Dole, therefore, opposed the April 2001 banana agreement, subjecting itself to criticism from labor unions and 170 NGOs that opposed FCFS. They accused Dole of "leading a race to the bottom in the banana industry" (US/LEAP 2001, 6) and favored the new banana agreement over the FCFS policy, which they feared would trigger a production shift toward Ecuador and away from Latin American countries whose banana labor forces were highly unionized and better paid.

With U.S. interests accommodated, EU efforts focused on Ecuador, which again proved to be an obstacle to achieving an agreement. It was the most positively disposed toward FCFS and was accustomed to pursuing an independent path on banana trade issues. Ecuador was pleased with the transfer of one hundred thousand tons from the ACP quota to TRQ B and favored the elimination of country quotas, so its initial response was not totally negative. But it was concerned about the new licensing arrangement and the ability of Ecuadorian firms classified as nontraditional operators to gain the coveted import licenses. Therefore, it did not support the new regime at first. This was a source of concern because, as the largest supplier of bananas to the EU, Ecuador qualified as a principal exporter, which allowed it to play an obstructionist role in the effort to gain WTO approval for the accord. Its favorable response, therefore, was critical to ending the dispute.

On April 30, 2001, the EU and Ecuador announced an accord that was compatible with the U.S.-EU agreement. The agreement detailed the provisions through which firms could qualify for the nontraditional operators' licenses, and the EU assured Ecuador that a significant share of licenses would go to nontraditional operators (EC 2001b, 2). The share ultimately set aside for new operators was 17 percent (EC 2001a, 2), very close to the percentage Ecuador had requested. Ecuador agreed to work to gain WTO acceptance of the new policy.

With Ecuador now on board, three major parties to the dispute were in agreement. This was not true, however, of all the lesser players. The Caribbean response to the new banana regime combined relief and disappointment. The demise of the FCFS plan was a source of relief because the plan had been viewed as a death knell for the Caribbean banana industry. The disappointment resulted from the reduction in the size of the ACP quota and the absence of individual country quotas (CBEA 2001, 1). There was disunity within CARICOM over the regime, however, with Suriname and Belize representing a minority view. The unique structure of their industries offered the opportunity to expand, given the right market conditions, and the smaller ACP quota would not be favorable to realizing this potential. Suriname opposed linking import license quantities to companies rather than to countries because it would place the countries at a disadvantage when negotiating with the corporation(s) handling their bananas. In Suriname and Belize, the firm was Fyffes, which also sourced in many other countries; so for them, a link between licenses and companies would mean less security for their producers. Suriname did not view the new accord as final (Sahtoe 2001, personal communication), and in August 2001, before the temporary demise of its industry, it pursued further discussions with the EU, Belize, and the ACP negotiating machinery in Brussels. The director of Surland played a major role in developing the strategy, but CARICOM was not involved because of the lack of consensus on the issue. Belize and Suriname hoped to convince the EU to add a new category of licenses within the ACP quota that would permit traditional ACP states to obtain licenses independently of any specific company. This would allow the two to market bananas directly in Europe without being tied to Fyffes. To facilitate this arrangement, a rolling reference period would be needed to determine the quantity of bananas for which such licenses would be granted (Bradley and Moriah 2001, personal communication).

There were three potential losers under the new policy. First, the former BFA exporters lost their guaranteed market shares. Second, the European import companies that had benefited from Council Regulation 404/93's licensing provisions for firms that traditionally handled ACP bananas would face reductions in their shares. With the right to handle 30 percent of the original third-country tariff quota of two million tons, many of the companies had improved their market shares over the course of the decade. The earlier historic reference period was not beneficial to them. This was a surprising result, considering that the EU had prolonged the dispute to enable those companies to establish footholds in dollar-zone producing states, but the new regime counteracted this progress. Third, on a smaller scale, fair trade companies that imported bananas into the EU were negatively affected for the same reason.[1] Their access to licenses declined because their increased share mostly occurred after the 1994–96 reference period used by the new policy for awarding licenses.

The WTO and the Banana Trade Dispute: A Key Test to a New System

The creation of the Single European Market that triggered the banana war occurred just before the WTO was established in 1995.

Several parties to the dispute that had not participated in the GATT 1947—notably, Ecuador—quickly acceded to the new organization after its founding. The WTO had barely begun to develop a track record of its own at the point when it was presented with U.S. and Latin American complaints about what would prove to be a very complex set of issues.

The banana trade dispute was a major test of the new WTO system. It represented a challenge on several fronts. First, the case involved a broad spectrum of the WTO membership, beyond the typical disputes between two countries. Second, the case directly related to several key articles within the GATT 1994 and was instrumental in setting precedents for how those articles were enforced. Third, the banana issue illustrated the potential effectiveness of the WTO's capability to invoke trade sanctions in sectors other than the one directly involved in the dispute. Fourth, the involvement of the United States, which was not an exporter of bananas, forced the system to consider whether a country could act as a complainant on behalf of corporations if its own exports were not involved. Finally, the case tested the new dispute settlement process. Together, these issues made the case one of the most important ever encountered by the new institution governing international trade.

H. E. Roberto Betancourt, Ecuador's WTO ambassador, considered the case to be a test for the WTO generally and for its DSB specifically. He suggested that the main lesson learned from the case was that, despite years of trying, compliance was not gained as of January 2001. Ecuador won three favorable WTO decisions but had not won compliance. For Betancourt, the banana case illustrated that the new system was not sufficiently equitable to permit a country like Ecuador to challenge "a trade superpower like the EU." The WTO authorized Ecuador to apply sanctions against the EU, but it did not do so. Betancourt said that it was difficult for countries like Ecuador to use sanctions such as tariffs to gain compliance because they would suffer greater economic consequences than the EU if they did so (Betancourt 2001, personal communication). This is an exceptionally critical point that extends beyond Ecuador to many Latin American countries and virtually the entire ACP group. Most LDCs depend on major economic powers for a variety of imports needed in their domestic industries. The tariffs involved in WTO-authorized sanctions would be paid by their domestic firms, which would make them less competitive on world and domestic markets—a result that would contradict the demands of the IMF and other forces that simultaneously drive the competitiveness imperative.

Jean-Jacques Bouflet, the EU's minister-counselor for legal affairs in Geneva, offered an interesting comment on the banana case, stating that it demonstrated an essential flaw in the WTO system. To Bouflet, the case illustrated how the process was negatively oriented—that is, it offered a way to prove the fault of one of the parties in a dispute but was less capable of deriving a positive solution to the problem. The DSB panel and the Appellate Body did not offer the answer; the party at fault had to devise its own solution. He also noted that there was no mechanism for ending sanctions once sanctions were authorized (Bouflet 2001, personal communication). In practice, an end to sanctions is negotiated as part of the agreement concluding a given dispute, as occurred with the banana case.

Dr. Kathy-Ann Brown, of CARICOM, suggested that the banana case tested the responsiveness of the new WTO system to the needs and concerns of "small players" like those she represented.[2] She perceived an inherent inequity in the dispute settlement process due to its marginalization of smaller countries on certain aspects of the process. As an example, she noted that the Caribbean countries were excluded from attending the hearings on the U.S. request to apply a "carousel" approach to its sanctions against the EU.[3] The United States requested the exclusion of the Caribbean countries, which was justified by the system's definition of the "major players" in each trade dispute case (Brown 2001, personal communication). Thus, while GATT 1994 seemed to recognize the participation rights of supplying countries whose export volumes are smaller but economically significant, participation did not occur on a consistent basis. Dr. Brown's comments added credence to the feelings expressed by many in the Caribbean that the combined effects of the creation of both the SEM and the WTO spelled double trouble for their region. That combination meant that individual Caribbean states could no longer achieve the critical threshold of 10 percent for any of their major exports in their primary markets—the minimum needed to gain a significant foothold in Geneva.

According to Sandiford (2000), the Caribbean's participation in the WTO will always be shaped by the ongoing power struggle between the United States and the EU. He noted the EU ban on imports of hormone-treated beef from the United States as one example of a dispute with implications that carried over to the banana issue and motivated the U.S. challenge to another EU policy. In such situations, the Caribbean is relatively powerless to prevent the two trading superpowers from engaging in those disputes.

Kym Anderson (2000, 18) suggests that the WTO dispute settlement system was not designed to deal with situations involving TNCs producing outside their home country. While the system allows member states to bring complaints against one another, in today's global economy, countries that are not exporters of a good often become involved in disputes on behalf of their TNCs. Anderson's view reflects the belief, widely held in the EU, that U.S. actions in the banana case were driven by Chiquita. This perspective was manifested in the EU's unsuccessful effort to get the WTO to decline to hear the U.S. complaint at all. The decision to hear the case has set a precedent for indirect TNC participation in the WTO's future affairs. Thus, although TNCs cannot be members of the WTO, their influence is secure because of the banana case decisions, which will likely add to public perceptions that the WTO is just an instrument of international capital.

The broader significance of the banana trade dispute to the WTO system is perhaps best summarized in the WTO's report on the case, a volume whose length exceeds one thousand pages. Therein, the DSB panel for the United States' complaint against the EU notes that the case was an exceedingly complex one, involving six parties (one representing fifteen member states) and twenty third parties, collectively almost one-third of all wto members. Claims were made under the GATT 1994 and, for the first time, under four other wto agreements: the Agreement on Agriculture, the Agreement on Import Licensing Procedures, the Agreement on Trade Related Investment Measures (TRIMS), and the GATS (wto 2000, 949).

After several years and so much effort, what was the result? In the final analysis, for the wto itself, the outcome was a positive one. Compliance, the preferred outcome, was achieved. The noncompliant policy was withdrawn, although sanctions were used in the interim. However, there were serious difficulties involved with the banana case. Numerous issues surfaced; some of those went unresolved and others resulted in problematic conclusions. The outcomes of the banana case are likely to be felt in subsequent complaints involving commodities and to influence trade policy in many wto member states during the early decades of the twenty-first century.

The End of the War:
a Neoliberal Triumph in Globalization's March?

The U.S.-EU banana trade dispute was a key event in a process of globalization that is extending further into the agricultural sector. While secondary and tertiary activities had already been through major transformations resulting from technological advances that fostered greater economic integration, proponents of globalization, such as TNCs, faced many obstacles and restrictions that limited their maneuvering space in the primary sector. In 1995, when the wto began operations, first world agriculture still had high protectionist barriers and subsidies, second world agriculture was in the process of transition from collectivization to privatization, and export agriculture in the developing world was dependent on historic trade agreements offering preferential access to first world markets.

Countries favoring increased globalization in agriculture looked to the wto to aid that process, and countries opposed to globalization feared what the wto would do.

The outcomes of the banana case constituted a giant step forward for those who promoted neoliberal economic policies in the agricultural sector. Several indicators support this argument, and some of them transcend the particulars of the wto decision itself. First, the decision to allow the United States to file a wto complaint involving a product it did not export was based on the fact that the United States was the host country of the tncs claiming to experience economic damage from the EU policy. Thus, the wto's rejection of the EU's claim that the United States, as a nonproducer, had no rights in the case represents a clear triumph for tncs around the world, including the EU-based firms that ostensibly lost in this case. Transnational corporations have led the globalization and neoliberalism processes since their inception, and because so many contemporary global trade flows occur *within* such firms, they are the primary beneficiaries of free trade. Nevertheless, the multilateral institutions responsible for managing the world economy are organizations to which states, not corporations, belong, leaving the tncs with no formal status in organizations such as the wto. The United States' victory in its request to be a complainant in the case represents an important precedent that will facilitate an increase in the very kind of corporate influence that the wto's many critics already target.

Second, at their most basic level, the wto rulings in favor of the United States and Ecuador represented a victory for the supporters of free trade over those countries (i.e., the EU and the acp group) favoring a managed trading structure that maintained historical preferences. Ironically, the United States itself maintains preferences for several selected agricultural imports and in 2002 passed a new farm subsidy bill. Nevertheless, the wto decisions do not bode well for the future of trade preferences, particularly quotas, at least where trade preferences are instituted in such a way as to restrict the economic opportunities available to other major players in a given industry.

Closely related to the victory of free trade is a third indicator of the expansion of the neoliberal economic agenda: adjustments to the Lomé system. By 1995 the Lomé Convention represented the largest economic relationship between developed and developing worlds. Despite its neocolonial nature, the Lomé system was staunchly supported by the ACP states because it provided them with a lucrative, guaranteed market for many goods. After 1996 the EU and the ACP had to negotiate the system's future against the backdrop of the ongoing banana trade dispute, and the influence of the dispute on the outcome of the negotiations is irrefutable. Despite substantial EU rhetoric to the contrary, Lomé's replacement, the Cotonou Agreement, is a significant retreat from the goals and methods of the Lomé arrangement. To its credit, Cotonou creatively sought new forms of cooperation. In June 2002 the EU endorsed a plan to negotiate Economic Partnership Agreements (EPAs) with seventy-six members of the ACP group during a six-year period ending in 2008. The EPAs extend the tariffication principle to nearly all exports from states entering the EU, but they also promote such neoliberal goals as regional integration and liberalization of ACP economies (EC 2002, 2).[4] Also, the quota preferences that were a Lomé trademark are passing into history, and duty-free entry of ACP goods into the EU does not eliminate the need to be competitive when tariffication alone proves insufficient to offset the ACP producers' competitive disadvantages.[5] This weakened EU-ACP cooperation system, then, represents another victory for free trade and the competitiveness imperative.

A fourth indicator of the triumph of neoliberalism is the fate of the licensing system employed by the EU in its 1993 banana policy. That system was effectively dismantled by the final agreement. The 1993 EU banana import licensing system was certainly complex, rather opaque to many observers, and easily criticized for the benefits it offered to EU firms. Nevertheless, the true purpose of the system was to provide incentives for EU-based TNCs to continue operations in uncompetitive ACP states. That it did so at the expense of dollar-zone companies was, of course, the deciding factor in bringing the

United States into the fray. But the ramifications of this particular aspect of the banana case are far-reaching for the Caribbean ACP states and, most likely, for African and Pacific ACP countries whose trading relations with the EU involve commodities other than bananas. A major export industry (e.g., bananas or sugar) carries with it an array of services and related industries, including transportation. The failure of such an industry, due to the removal of preferences and a protective licensing system, will have severe repercussions throughout the nation's economy. This is very true of the smallest countries, particularly the islands that are well represented in the ACP group. Should bananas fail in the eastern Caribbean, the ships necessary to bring Dominica's imports or to carry Grenada's other exports to Europe might cease to call.

The marginalization of the small ACP states involved in the dispute is a fifth indicator of the neoliberal victory in the banana war. With the emergence of integrated trading blocks, small countries find their ability to influence events in the global trade regime, which was never great, becoming ever more diminished. The EU, with its SEM, has led the way toward economic integration, but other groups now follow in its path. This places many small economies in an untenable situation from which it will be difficult to escape. The banana case illustrates the problem for small states that might be significant exporters of a good to individual EU members but are not very important to the EU supply overall. The substantial supplier stipulations precluded the meaningful participation of ACP states in the negotiations that determined their future ability to export bananas to their major market. The inability of the Caribbean states to exercise substantial influence on the outcome of the banana case suggests that, in the future, other small states will be marginalized in such processes. It is unlikely that countries like Dominica and Grenada will ever become substantial suppliers of goods to large collective markets. Nevertheless, it is that collectivity that is now subject to WTO trade rules.

By early 2004 it was already evident that the Windward Islands

TABLE 8. Banana farmers in Windward Islands

Country	Number of farmers	
	In 1992	*In 2001*
Dominica	5,800	1,300
St. Lucia	9,700	3,800
St. Vincent	7,800	2,200

Source: NERA 2003, P.28

had experienced severe declines in their banana industries. St. Lucia's banana exports sank to just 32,520 tons in 2003, from 122,066 tons in 1992, the year before the new EU policy went into effect. Corresponding figures for Dominica were 10,823 tons in 2003, down from 51,606 in 1992; and for St. Vincent, 20,919 tons in 2003, down from 71,320 (Technical Centre for Agricultural and Rural Cooperation 2006). The number of banana farmers declined precipitously as well, as indicated in table 8.

This situation is extremely frustrating to government officials in small states, who argue for "trade, not aid." They are really advocating fair trade, not the free trade that is the goal of the WTO. Their argument in favor of fair trade emanates from their recognition of the natural competitive limitations that would surely render them among the losers in virtually any completely free trade situation. Nevertheless, the neoliberal ascension in major international forums makes it increasingly difficult for such countries to influence the rules of trade. The outcomes for many will be their greater dependency on whatever foreign aid the North may wish to give and increased social and political instability that may threaten democratic governments.

Labor was deeply affected by the banana dispute, and the implications of the agreement for workers represent a sixth indicator of the neoliberal victory. Like the fruit TNCs, labor organizations had no direct participation rights in WTO deliberations, and the Latin

American governments involved in the dispute exhibited little interest in representing labor's cause. The declining profitability of the banana industry during the dispute exacerbated labor's problems. The TNCs operating in Latin America reacted as companies in many other industries have during the past twenty years—with layoffs, reduced real wages, and an increased reliance on subcontractors, all of which weakened labor's position relative to management in the industry. In the eastern Caribbean, employment in the industry declined substantially, falling below thirty-six thousand by 1998, with estimates of even more serious declines with full trade liberalization for bananas (Sandiford 2000, 117–23).

Finally, although Ecuador would ultimately negotiate an acceptable settlement, its hollow victory in its WTO case is perhaps the most disturbing indicator of the neoliberal triumph. Although Ecuador gained a positive decision in its WTO battle, it found that it could not win the war. The process reaffirmed the most essential reality of the world economy since the early colonial era: the virtual impossibility of successfully challenging the global economic power structure. For Ecuador, applying the authorized sanctions would have done greater harm to its economy than to the EU's. As H. E. Roberto Betancourt noted, Ecuador's experience with the case illustrated the basic inequity of the WTO system and the inherent advantage it affords to the more powerful economies. This reality could influence LDCs' future decisions about whether or not to challenge the economic powers within the WTO.

The Unfinished Agenda of the Banana War

Although the U.S.-EU banana war apparently reached its conclusion, it left many issues unresolved, including several issues related to the future of the EU, the source of the controversial policy that led to the dispute. What did the EU learn from the experience that will affect how it conducts its affairs in the future? Will it continue to pursue the contradictory priorities that led to the development of a banana importation policy that was virtually guaranteed to create problems? Is it now aware of the limited ability of trade preferences to improve the economies of LDCS? The EU's eastward expansion to twenty-seven members in 2004 and 2007 is diluting its resources, causing it to reevaluate its priorities. Will that process affect its relations with the ACP group and with Latin America?

The future role of the WTO and its real impact in the developing world also remain unresolved. Although the Doha Agreement of 2001 represents an attempt to ensure that the interests of LDCS are better accommodated, Doha Round negotiations were seriously derailed in 2003 when major LDCS walked out of the ministerial conference in Cancún, Mexico, over the refusal of developed states to negotiate their agricultural subsidies. If Doha fails, will developing and underdeveloped states remain committed to the WTO or will they see it as irrelevant, like the 1947 GATT? Will there ever be a time when a country such as Ecuador or Dominica can successfully challenge the EU, the United States, or Japan? Will WTO rules be modified to protect the rights of small countries that do not qualify as substantial suppliers of a given good to participate in dispute negotiations?

The future of the ACP group also merits further attention. The Caribbean states in the group appear to be the real losers of the banana war. The end of preferences arrived in 2006, and with continuing involvement of the WTO, the EU implemented a new, compliant banana importation policy that features tariffication. As feared by the Caribbean ACP exporters, the tariff was set at just 176 euros per ton, not high enough to offset their competitive disadvantage. The EU will maintain a special duty-free ACP quota of 775,000 tons, but this will not have individual country allocations and it will be available on a first-come, first-served basis that favors those ACP states—primarily African countries, but also Belize and Suriname—that have the ability to expand their production levels. Once the quota has been filled each year, the 176 euro tariff will be applied (CBEA 2006). Thus, an outcome in which the Windward Islands will be able to regain the export production that they have already lost appears unlikely. What will happen in the Caribbean now that the new rules have been implemented? What kind of flexibility will be shown to help those still unable to resolve economic competitiveness shortcomings, whether in bananas or other goods? Will the failure of a given country's banana industry ripple through its national economy, affecting other sectors as well? The role of the Cotonou Agreement and its new EPAS merits research to ascertain whether or not the agreement is producing the desired results.

Finally, the triumph of free-market principles bodes well for banana industries in Latin American. The predictions of doom that were made there in 1993 have mostly not been realized. But some problems remain. Can the public sectors of Latin American banana-exporting countries recover their losses and resume the march forward that began in 1974? Will unionized labor be able to reassert itself in Latin America's industries to recover the ground it lost after 1993? Will Ecuador improve workers' rights, wages, and living standards to end the race to the bottom?

Ultimately, the outcomes of the banana dispute indicate that the competitiveness imperative reigns and will continue to cause problems for uncompetitive nations. Barbara Welch (1996, 322), writing about the eastern Caribbean, offered this perceptive encapsulation of this dilemma: "In the Lesser Antilles, there is nothing that cannot be grown—but there is nothing that cannot be grown more cheaply somewhere else." Her words certainly ring true for the governments of small developing states around the world as they ponder which development policies to pursue among what often seems to be a declining selection of options. Will niche markets in agriculture provide a stable, sustainable path? Or, as Oliver Benoit suggested, will it merely be a situation of continually "putting out fires," as countries exploit one niche after another in a search for the right goods to export to fickle markets (Benoit 1997, personal communication)? Does agriculture itself hold the answer for LDCs, given the difficulty of attracting investment to the sector? The only things that are clear are that these questions will not be answered easily and that great uncertainty remains about whether there will be a place at globalization's table for the world's small, uncompetitive countries.

The end of ACP trade preferences in 2006 rendered the EU's Council Regulation 404/93 a candidate for the archives of historical economic geography, alongside Charles Kepner and Jay Soothill's seminal 1935 work, *The Banana Empire: A Case Study of Economic Imperialism.* Efforts made in the 1990s to restructure ACP banana industries to make them more competitive and the energy devoted to diversify ACP economies have yielded few solid results thus far. Little time remains to resolve these problems.

Although there is little rational cause for optimism about the outcome, there are some encouraging signs in the search for answers. Interest in the plight of uncompetitive countries, particularly the small ones, has increased in the WTO, the UNCTAD, and the Commonwealth of Nations. It is quite clear that agriculture alone, with its many complex problems, is not likely to provide the needed solutions. The solutions may well lie in the secondary or, more probably, in the tertiary and quaternary sectors of the global economy, particularly in activities that are relatively aspatial. Such activities could reduce obstacles such as distance to markets, transportation costs, and the inability to achieve economies of scale. Possible activities include informatics and telecommunications-dependent back-office service operations, which do not involve the physical movement of products. These activities, by their very nature, are flexible with regard to location, which can also make them unreliable in the long run because they can be abandoned with relative ease if more lucrative opportunities are presented elsewhere.

These issues will provide many academicians, government functionaries, and private-sector social scientists with an active research agenda in the years ahead. I hope to be among them, moving beyond bananas to search for the answers to these important questions.

Notes

Introduction

1. The terms *third country* and *dollar zone* are drawn directly from EU documents.

1. The Creation of the Banana Empire

1. Many of the Jamaicans remained in Costa Rica, and their descendants continue to live in the Puerto Limón area. They were finally granted citizenship during the 1950s.

2. Ellis (1983, 44) describes a 1906 United Fruit Company purchase of 50 percent of the shares in the Vaccaro Brothers Company, one of its principal potential rivals in Honduras. The U.S. court system thwarted this acquisition as a violation of antimonopoly legislation, so the company sold its shares in 1908. Ironically, the Vaccaro Brothers Company was reorganized in 1924 as the Standard Fruit Company, subsequently known as the Standard Fruit and Steamship Company, and became the chief competitor to the United in Central America.

2. The Empire Challenged

1. "Nine-hand bunch" refers to the output of one plant, measured vertically before being cut into the smaller bunches familiar to consumers. A nine-hand bunch was the industry's standard measurement.

2. By the 1980s the Urubá zone was manifesting the same characteristics of decline, and the TNCs operating in the region began to move once again, this time returning to the Santa Marta area. This second shift threatened the livelihood of at least twenty thousand workers directly or indirectly employed in the industry in Urubá, the forty thousand people dependent upon them, and the six thousand homes provided by the companies for those families (Sierra 1986, 11–16).

3. Kepner and Soothill (1935, 137) describe one instance in Honduras in 1932 in which the government declared martial law and sent in troops to break a strike declared by workers of the Tela Railroad Company. This enabled the UFCO to force the strikers to accept a reduction in wages.

4. Indeed, many Costa Ricans consider the book *Mamita Unai* (literally "Mother United," in reference to the UFCO), which focuses on that strike, to be the country's national novel.

5. Following Costa Rica's short civil war of 1948 and the abolition of its army the following year, the National Liberation Party (Partido de Liberación Nacional, or PLN) dominated the country's politics. While remaining socially progressive and maintaining Costa Rica's social security system, it was fiercely anticommunist. The PLN pursued interventionist policies where labor strife was concerned, driving the leftist elements within labor organizations underground, except in the banana zones, where they continued to confront the TNCs.

6. Of the fifty-five thousand acres, ninety-four hundred were cultivated in bananas at the time of purchase.

3. The End of Splendid Isolation

1. While the tax increases may seem rather high, the boxes involved weighed forty pounds, rendering the tax a more reasonable one cent to two and one-half cents per pound.

2. In Guatemala's case, the decision not to implement a tax at that time was based on the fact that doing so would violate the country's contract with Del Monte, which extended to 1981.

3. Among these were the Corporación Bananera Nacional (CORBANA) in Costa Rica, the Oficina Nacional del Banano (later the Dirección Nacional) in Panama, and the Corporación Hondureña del Banano (COHBANA) in Honduras.

4. This situation was described at length by Antonio Montero, of the Instituto Centroamericano de Asesoría Laboral, and was documented in an unpublished environmental study conducted by the International Union for the Conservation of Nature.

5. During my visits to Costa Rica in 1989 and 1990, Costa Ricans in the rest of the country still associated the Golfito area with images of poverty and massive unemployment. The government's response to the problem

was to open a duty-free shopping zone in the town in 1990, allowing Costa Ricans to avoid the country's still high tariffs by traveling to Golfito. Many did so, despite the relative inaccessibility of the region. I saw long lines of people laden with electronics equipment and other purchases when I was there in July 1990.

4. Peasant Farmer Societies

1. Among these commissions were the British West Indian Royal Commission of 1897 and the Moyne Commission of 1939, both of which advocated improving access to land for many of the region's small-scale farmers, among other recommendations. The latter also recommended an emphasis on mixed farming of food crops instead of export production (Brierley 1996, 2).

2. The use of the word *peasant* is quite common in the literature on small-scale, independent farmers of the eastern Caribbean. It is also heard in everyday conversation in the region itself.

3. Geest entered the British market as a supplier of flower bulbs and garden products. By the early 1950s it already had a distribution system in place in the UK, which proved advantageous when it began marketing bananas there.

4. "Dollar zone" is the EU term referring to the Latin American banana-producing zone. The implication, not altogether inaccurate, is that the region's banana industry is controlled by major U.S.-based fruit transnationals. The term is frequently used in EU documents.

5. Dominica operated for many years with two closely affiliated organizations. The Dominica Banana Growers Association represented the interests of its banana farmer members while the Dominica Banana Marketing Corporation handled many of the administrative functions. The two organizations merged in 1995 under the name of the Dominica Banana Marketing Corporation (Durand 1996, personal communication).

6. Boxes are produced on St. Lucia and are supplied from there to the other three islands. The DGBA's Conrad Cyrus described prices as high, indicating another economies of scale issue.

7. Trouillot (1988, 41) describes small markets involving hucksters and other independent traders in the export of rejected Dominican bananas within the region.

5. Belize, Suriname, and the French West Indies

1. Creole, the language of Afro-Belizeans, is spoken at home by 40 percent of the population. Three Mayan languages, Garífuna, and German (spoken by Mennonites) are also important. Most people learn English at school because it is the language of instruction—except, significantly, in the banana zone, where Spanish is used.

2. See Moberg (1997) for an exhaustive discussion of labor issues in Belize's banana industry.

3. Ironically, the cartons are imported from Mississippi, in the United States, which is seen as the country determined to destroy Belize's banana industry. Fertilizers for bananas are also purchased from the United States.

6. The Single European Market

1. The ECU, or European Currency Unit, preceded the euro. Although it was never issued in the form of coins or notes to be spent by consumers, it did have trading value on world currency markets and was used as the official currency of the EEC/EU in its documents. Its value at the point when Council Regulation 404/93 went into effect was approximately $1.20.

7. Neocolonialism Encounters

1. The EDF was and continues to be programmed for five-year periods. Each period is, therefore, named for an ordinal number corresponding to its place in the chronology of the program (i.e., first EDF, second EDF, and so on).

2. The waiver extends until 2008, at which point reciprocity would begin (Josling 2001, 184).

8. The World Trade Organization

1. The agreements include the Marrakech Agreement Establishing the World Trade Organization and fifteen other agreements that cover various trade sectors and mechanics of governing trade activities.

2. The Cairns Group includes Australia, Canada, New Zealand, and four-

teen developing countries—Argentina, Bolivia, Brazil, Colombia, Costa Rica, Fiji, Guatemala, Indonesia, Malaysia, Paraguay, the Philippines, South Africa, Thailand, and Uruguay. It continues to meet to ensure that the WTO Agreement on Agriculture is implemented properly and expeditiously.

9. The U.S.–EU Banana War

1. At that time, neither Ecuador nor Panama was a signatory power to the GATT, so they could not file complaints.

2. The four Windwards paid twenty million pounds in cash. Fyffes also paid twenty million. The remainder was financed through loans, using the company's physical assets as collateral.

3. Quota A included 2.2 million tons, reflecting the third-country imports into the twelve EU member states as of 1993. Quota B included an additional 353,000 tons for the three countries that were not members when the initial policy was implemented.

10. Pursuit of an Elusive Goal

1. "Fair trade," as used here, refers to getting a fair deal in trading relationships. It can also refer to a trade system intended to ensure that farmers or artisans receive a fair price for their goods.

11. Implications for the Future

1. Fair trade companies are often connected to NGOs that promote payment of higher-than-market prices for agricultural commodities, crafts, and other goods imported from developing countries. They attempt to alleviate problems caused by fluctuations in market prices that sometimes carry prices below production cost levels, as is currently the situation worldwide with coffee.

2. Since January 2001, when I interviewed Dr. Brown in Geneva, the WTO has undertaken an effort to make the organization more responsive to the concerns of "small economies," potentially the same group of nations reflected in her comments. This was an outcome of the Doha Ministerial Conference that took place in Qatar later that year.

3. The carousel approach involved a rotating series of imports from the EU upon which the United States proposed to levy new tariffs. The purpose of rotation was to prevent market conditions from adjusting to the new tariffs on any particular group of goods, thus allowing the pressure inherent in the sanctions to continue to have the intended negative effect in EU member countries.

4. The extension of the tariffication principle is not quite the major step it appeared to be on the surface. By 2002, 93 percent of the exports from the ACP could already enter the EU duty-free.

5. As with bananas, the amount of the tariff preference for such ACP products will determine whether or not the strategy actually benefits those states. With the current trend toward tariff reduction now affecting agriculture more than in the past, it seems unlikely that tariffication alone will be sufficient to ensure access.

References

Published Works

ACP Secretariat. 1995. *Agreement Amending the Fourth ACP-EC Convention of Lomé*. Signed in Mauritius, November 4. http://ec.europa.eu/development/Geographical/Cotonou/LomeGen/LomeItoIV.cfm.

———. 2000. *The Cotonou Agreement*. Signed in Cotonou, Benin, June 23. http://ec.europa.eu/development/Geographical/CotonouIntro_en.cfm.

Agro Press. 1998. "Will the Banana War Take Place After All?" *Fruit World: The Journal for the International Produce Trade* 57 (3): 306–12.

Anderson, Kym. 2000. "The Future Agenda of the WTO." In *From GATT to the WTO: The Multilateral Trading System in the New Millennium*. Prepared by the WTO Secretariat. The Hague, Netherlands: Kluwer Law International.

Asociación Ecologista Costarricense. 1992. "Declaración de Casa Emaus: ALTO a la expansión bananera incontrolada." *El Ecologista* 2 (15–16): 6–10.

BananaLink. "Noboa." 2006. Norfolk, UK: BananaLink. http://www.bananalink.org.uk/index.php?option=com_content&task=view&id=66&Itemid=26&lang=en.

Barry, Tom, and Dylan Vernon. 1995. *Inside Belize*. Albuquerque NM: Interhemispheric Resource Center Press.

BBC News. 1999a. "Banana War Exposes Old Trade Divisions." March 5. http://news.bbc.co.uk/hi/english/business/the_economy/newsid_290000/290981.stm.

———. 1999b. "Europe Will 'Comply' with Banana Ruling," April 7. http://news.bbc.co.uk/hi/english/business/the_economy/newsid_313000/313362.stm.

Benns, J. A., J. R. Webb, and R. H. Stover. 1981. "Belize: A Reappraisal of the Banana Industry." A Fyffes Group Ltd.–United Fruit Company study of the industry. Dublin: Fyffes-UFCO.

Brierley, John. 1996. "Changing Farms Systems: From Estate Holdings to Model Farms, Gleanings from Grenada 1940–1992." Winnipeg, Manitoba: University of Manitoba.

Brittan, Sir Leon. 1995. "Letter to Michael Kantor, U.S. Trade Representative." Brussels, Belgium: European Commission, January 10.

Burbach, Roger, and Patricia Flynn. 1980. *Agribusiness in the Americas*. New York: Monthly Review Press.

Burton, Richard D. E. 1995. "The French West Indies *a le heure de l'Europe*: An Overview." In *French and West Indian: Martinique, Guadeloupe, and French Guiana Today*, edited by Richard D. E. Burton and Fred Reno. London: MacMillan.

Cairns Group. 2000. "The Cairns Group: An Introduction." Canberra, Australia: Commonwealth of Australia. http://www.cairnsgroup.org/introduction.html.

Carbaugh, Robert J. 2000. *International Economics*. Cincinnati: Southwestern Publishers.

Caribbean Banana Exporters Association. 1999. "CBEA Campaign: The Current Crisis." London: CBEA. http://www.cbea.org/cbea2/CBEA/action.htm.

———. 2000. "Annual Report for 1999/2000." London: London Lobby of CBEA.

———. 2001. "Banana Settlement Welcomed by Caribbean Industry." London: CBEA. http://www.cbea.org/PRESS/rd_18.htm.

———. 2006. "New EU Regime Threatens Most Vulnerable Banana Producers." London: CBEA.

CARICOM. 1996. "Transformation of Agriculture: Caribbean Community." Georgetown, Guyana: CARICOM Secretariat, June 14.

CBEA. See Caribbean Banana Exporters Association.

Center for Programs and Investment Projects. 1995. *Belize Agricultural Sector Study*. San José, Costa Rica: IICA.

Central Statistics Office (CSO). 1995. *Belize External Trade Bulletin 1994*. Belmopan, Belize: Ministry of Finance, May.

CEPAL. See Economic Commission for Latin America and the Caribbean.

CEPPI. See Center for Programs and Investment Projects.

CORBANA. See Corporación Bananera Nacional.

Corporación Bananera Nacional. 1993a. *Informe anual de estadísticas de exportación de banano*. San José, Costa Rica: CORBANA.

————. 1993b. *Repercusiones para Costa Rica de la aplicación de cuotas en el mercado bananero de la C. E.* San José, Costa Rica: CORBANA.

Dosal, Paul J. 1993. *Doing Business with the Dictators: A Political History of United Fruit in Guatemala 1899–1944.* Wilmington DE: Scholarly Resources Books.

Economic Commission for Latin America and the Caribbean. 1979. *Transnational Corporations in the Banana Industry of Central America.* New York: UN ECOSOC.

Ellis, Frank. 1983. *Las transnacionales del banano en Centroamerica.* San José, Costa Rica: Editorial Universitaria Centroamericana.

European Commission. 1991. *The Community's Banana Market in the Run-up to 1993.* Brussels, Belgium: EC Section for Agriculture and Fisheries.

————. 1993a. "Bananas: The Real Issue Is the Survival of Our Country." *Courier: The Magazine of ACP/EU Development Cooperation* 140 (July–August): 19–25.

————. 1993b. "Commission Regulation (EEC) No. 1442/93 Laying Down Detailed Rules for the Application of the Arrangements for Importing Bananas into the Community." *Official Journal of the European Communities.* Brussels, Belgium: June 10.

————. 1993c. "Commission Regulation (EEC) 1443/93 on Transitional Measures for the Application of the Arrangements for Importing Bananas into the Community in 1993." *Official Journal of the European Communities.* Brussels, Belgium: June 10.

————. 1994. "Report on the EC Banana Regime." Brussels, Belgium: EC Directorate General VI.

————. 1995. "Country Report: Suriname." *Courier: The Magazine of ACP/EU Development Cooperation* 151 (May–June): 11–42.

————. 1996. "Lomé IV Convention as Revised by the Agreement Signed in Mauritius on 4 November 1995." *Courier: The Magazine of ACP/EU Development Cooperation* 155 (January–February): 3–25.

————. 1999. "Commission Proposes to Modify the EU's Banana Regime." Brussels, Belgium: EC press release. November 10.

————. 2000a. "Dossier: The New ACP-EU Agreement." *Courier: The Magazine of ACP/EU Development Cooperation* 181 (June–July): 3–25.

————. 2000b. "Overview of the Agreement 2000." http://ec.europa.eu/development/Geographical/Cotonou/Cotonou2000_en.htm.

European Community (EC). 2001a. "Commission Implements Regulation to Bring WTO Banana Dispute to an End." IP/01/628. Brussels, Belgium: May 2.

———. 2001b. "EU and Ecuador Reach Agreement to Resolve WTO Banana Dispute." IP/01/627. Brussels, Belgium: April 30.

———. 2001c. "The EU Import Regime and the WTO Findings." Memo/01/135. Brussels, Belgium: April 11.

———. 2001d. "U.S. Government and European Commission Reach Agreement to Resolve Long-standing Banana Dispute." IP/01/562. Brussels, Belgium: April 11.

———. 2002. "EU Adopts New Strategy for Trade Relations with African, Caribbean, and Pacific Countries." IP/02/871. Brussels, Belgium: June 17.

European Council. 1993. "Council Regulation (EEC) No. 404/93 on the Common Organization of the Market in Bananas." *Official Journal of the European Communities*. Brussels, Belgium: February 13.

———. 2001. "Council Regulation (EC) No. 216/2001 Amending Regulation (EEC) No. 404/93 on the Common Organization of the Market in Bananas." *Official Journal of the European Communities*. Brussels, Belgium: February 2.

European Parliament. 1998. "Draft Report on the Legislative Proposal for a Council Regulation Amending Regulation (EEC) No. 404/93 on the Common Organization of the Market in Bananas." Strasbourg, France: Committee on Agriculture and Rural Development.

European Report. 1993. "EC Court Rejects German Attempt to Block New Regime." No. 1871. Brussels, Belgium: European Information Service, June 30.

European Union (EU). 2000. *The Lomé Convention*. http://europa.eu.int/comm/development/cotonou/lome_history_en.htm.

FAO. See Food and Drug Organization.

Financial Times (London). 1994. "Caribbean Slips Up over EU's Banana Regime." February 6.

Food and Agriculture Organization. 1999. "Exámen de los últimos acontecimientos en materia de políticas que influyen en el comercio del banano." Gold Coast, Australia: FAO Committee on Problems of Basic Products.

———. 2003. *The World Banana Economy 1985–2002*. Prepared by Pedro Arias, Cora Dankers, Pascal Liu, and Paul Pilkauskas. Rome, Italy: FAO of the United Nations.

Gabrielli, Rolando. 1988. "Itinerario y acciones de un protagonista de la Guerra del Banano." *Analisis: Revista económica de Panamá y CentroAmérica* 10 (3): 4–13.

Garelli, Stéphane. 2006. "Competitiveness of Nations: The Fundamentals." In *IMD World Competitiveness Yearbook*. Lausanne, Switzerland: IMD. http://www.imd.ch/research/centers/wcc/competitiveness_funda mentals.cfm.

Government of Costa Rica. 1884/1989. "El Contrato Soto-Keith sobre deuda exterior y el ferrocarril." Excerpts reprinted in *The Costa Rica Reader*, edited by Marc Edelman and Joanne Kenen. New York: Grove Wiedenfeld.

Government of Ecuador. 1993. *Impacto de las restricciones a las importaciones de banano de la Comunidad Europea para el Ecuador*. Quito, Ecuador: Ministerio de Agricultura y Ganadería.

Grossman, Lawrence. 1993. "The Political Ecology of Banana Exports and Local Food Production in St. Vincent, Eastern Caribbean." *Annals of the Association of American Geographers* 83 (2): 347–67.

———. 1998. *The Political Ecology of Bananas: Contract Farming, Peasants, and Agrarian Change in the Eastern Caribbean*. Chapel Hill: University of North Carolina Press.

Guardian (London). 2002. "Unfair Trade Winds." http://www.guardian .co.uk/food/focus/story/0,,956649,00.html.

Hayes, Rita Derrick. 1999. "U.S. Request to WTO for Tariff Authorization in Banana Case." Geneva, Switzerland: U.S. Permanent Representative to the WTO.

Heusnet, William. 1995. "Report: Cowpen 3 and 4." Belize City, Belize: Human Rights Commission of Belize.

Hintjens, Helen. 1995. "Constitutional and Political Change in the French Caribbean." In *French and West Indian: Martinique, Guadeloupe, and French Guiana Today*, edited by Richard D. E. Burton and Fred Reno. London: MacMillan.

Hirsch, Bruce. 1998. "The WTO Bananas Decision: Cutting through the Thicket." *Leiden Journal of International Law* 11 (2): 201–27.

Houston Chronicle. 2004. "Ruled by Fear, Banana Workers Resist Unions." January 19. http://www.chron.com/disp/story.mpl/special/04/leftbe hind/2095828.html.

Inter-American Institute for Cooperation on Agriculture (IICA). 1996. "Technological Modernization of the Banana Industry in the Caribbean." San José, Costa Rica: IICA Regional Projects, Planning, and Programming Unit.

International Institute for Management and Development (IMD). 2000. *World Competitiveness Yearbook*. Lausanne, Switzerland: IMD.

———. 2003. *World Competitiveness Yearbook*. Lausanne, Switzerland: IMD.

International Union for the Conservation of Nature (IUCN). 1992. *Diagnóstico del impacto socioambiental de la expansión bananera en Sarapaquí, Tortugero, y Talamanca, Costa Rica*. San José, Costa Rica: IUCN.

Jackson, John H. 2000. "Dispute Settlement and the WTO: Emerging Problems." In *From GATT to the WTO: The Multilateral Trading System in the New Millennium*. Prepared by the WTO Secretariat. The Hague, Netherlands: Kluwer Law International.

Jos, Emmanuel. 1995. "The Declaration of the Treaty of Maastricht on the Ultra-peripheral Regions of the Community: An Assessment." In *French and West Indian: Martinique, Guadeloupe, and French Guiana Today*, edited by Richard D. E. Burton and Fred Reno. London: MacMillan.

Josling, Tim. 2001. "Regional Trade Agreement and Agriculture: A Post-Seattle Assessment." In *Globalization and Agricultural Trade Policy*, edited by Hans J. Michelmann, James Rude, Jack Stabler, and Gary Storey. Boulder CO: Lynne Rienner Publishers.

Journal of Commerce. 1999. "EU Report on Banana War Options Highlights Complexity of Dispute." May 28.

Kantor, Michael. 1995. "Letter to The Honorable Sir Leon Brittan, Commissioner of the European Community." Washington DC: USTR, January 9.

Kepner, Charles D. 1936. *Social Aspects of the Banana Industry*. New York: Columbia University Press.

Kepner, Charles D., and Jay Soothill. 1935. *The Banana Empire: A Case Study of Economic Imperialism*. New York: Vanguard Press.

Lamy, Pascal. 2001. "The Commission and the US Administration Have Agreed on a Solution to the Banana Dispute." Statement issued in Brussels, Belgium, April 11.

La Nación (San José, Costa Rica). 1994a. "Nicas a merced de contratistas." January 17.

————. 1994b. "País frena a ilegales." January 18.

Langley, Lester D., and Thomas Schoonover. 1995. *The Banana Men: American Mercenaries and Entrepreneurs in Central America, 1880–1930.* Lexington: University of Kentucky Press.

Marie, J. M. 1979. *Agricultural Diversification in a Small Economy—The Case for Dominica.* Cave Hill, Barbados: University of West Indies Institute of Social and Economic Research.

McMahon, Joseph A. 1998. "The EC Banana Regime, the WTO Rulings, and the ACP: Fighting for Economic Survival?" *Journal of World Trade* 32 (4): 101–14.

Moberg, Mark. 1997. *Myths of Ethnicity and Nation: Immigration, Work, and Identity in the Belize Banana Industry.* Knoxville: University of Tennessee Press.

National Economic Research Institute (NERA). 2003. "Banana Exports from the Caribbean since 1992." London: NERA.

New Chronicle (Roseau, Dominica). 1994a. "An Attempt to Overturn Banana Protocol." December 2.

————. 1994b. "Caribbean Calls for Approval of GATT Waiver." December 2.

Newsome, Justine, and James Wilson. 1999. *Latin American Banana Growers Hail WTO Ruling.* London: Reuters EU Briefing, April 9.

Oficina Nacional del Banano. 1975. *Memoria Annual.* Panama City, Panama: Ministerio de Comercio e Industria.

Organization of American States (OAS). 1975. *Sectoral Study of Transnational Enterprise in Latin America: The Banana Industry.* Panama City, Panama: OAS Executive Secretariat for Economic and Social Affairs and the Oficina Nacional del Banano.

Paredes, Virgilio. 1976. "Significación de los productores y exportadores nacionales de banano en América Latina." Paper presented at the conference "Políticas de defensa de los productos de agricultura tropical: Problemas de la producción y comercialización del banano, la UPEB, y el comercio mundial." Comayagüeña, Honduras, April.

Parsons, James. 1957. "Bananas in Ecuador: A New Chapter in the History of Tropical Agriculture." *Economic Geography* 33 (3): 201–16.

Perillo, Robert. 2000. "Banana Workers and Transnationals: An Industry in Crisis." Chicago: U.S./Labor Education in the Americas Project.

Peter, E. R. 1982. *Limited Survey of the Banana Industry in the United Kingdom.* Belize City, Belize: Government of Belize.

Petersmann, Ernst-Ulrich. 1999. "The WTO Panel and Arbitration Reports on the EC Banana Regime." *Bridges* 3 (3): 3–4.

Reuters. 1993a. "Belgium Backs Germany in EC Banana Deal Challenge." June 11.

———. 1993b. "EC Set to Clamp Down on Banana Imports as Germany Loses Case." June 29.

———. 1993c. "France Says Banana Ruling Good for Caribbean." June 30.

———. 1993d. "GATT Sets Up Panel on New EC Banana Regime." June 16.

———. 1993e. "Germany Challenges New EC Banana Regulations." May 17.

———. 1998. "The U.S./E.U. Banana Dispute: Unilateral U.S. Retaliation Not in Line with WTO Rules and Politically Unwise, Sir Leon Brittan Says." November 10.

———. 1999. "European Exporters Confused by War of Words." April 8.

Reynolds, Philip Keep. 1927. *The Banana, Its History, Cultivation, and Place among Staple Foods.* New York: Houghton Mifflin.

Sandiford, Wayne. 2000. *On the Brink of Decline: Bananas in the Windward Islands.* St. Georges, Grenada: Fedon Books.

Secretaría Técnica Conjunta. 2000a. "Commission Proposes Solution to End Banana Dispute." Brussels, Belgium: IICA/FAO, October 4.

———. 2000b. "U.S. Rejects New Commission Banana Proposal in Advance of Council." Brussels, Belgium: IICA/FAO, October 6.

Shoman, Assad. 1987. "Review of the Banana Industry in Belize." Paper presented to the Seminario Internacional A-14: La Economia Bananera y las Organizaciones de los Trabajadores en la Decada de los 80 en America Latina. Santa Barbara, Heredia, Costa Rica.

SIDSNet. 2000. "About Small Island Developing States (SIDS)." United Nations Division for Sustainable Development.

Siebert, Horst. 2000. "What Does Globalization Mean for the World Trading System?" In *From GATT to the WTO: The Multilateral Trading System in the New Millennium.* Prepared by the WTO Secretariat. The Hague, Netherlands: Kluwer Law International.

Sierra, Diego Miguel. 1986. "Urubá: Banano y Bienestar Social." Paper presented to the First Banana Congress, Medellín, Colombia, November 20.

Sizer, Nigel, and Richard Rice. 1995. *Backs to the Wall in Suriname: Forest Policy in a Country in Crisis.* Washington DC: World Resources Institute.

Soto Ballestero, Moisés. 1992. *Bananos: Cultivo y comercialización.* San José, Costa Rica: Litografía e Imprenta.

Technical Centre for Agricultural and Rural Cooperation, ACP-EU. 2006. "Banana: Executive Brief." *Agritrade*, July 9. http://agritrade.cta.int/en/content/view/full/1266.

Tico Times (San José, Costa Rica). 1993. "Limón Banana Farms Face Labor Charges." March 19.

Thomson, Robert. 1987. *Green Gold: Bananas and Dependency in the Eastern Caribbean.* London: Latin American Bureau.

Thrupp, Lori Ann. 1995. *Bittersweet Harvests for Global Supermarkets: Challenges in Latin America's Agricultural Export Boom.* Washington DC: World Resources Institute.

Trouillot, Michel-Rolph. 1988. *Peasants and Capital: Dominica in the World Economy.* Baltimore: Johns Hopkins University Press.

Union of Banana Exporting Countries (UPEB). 1974. *Convenio Constitutivo de la Unión de los Países Exportadores de Banano.* Panama City, Panama: UPEB.

———. "La Actividad Bananera 1992." 1993a. In *Informe UPEB*, no. 95. Panama City, Panama: UPEB.

———. 1993b. "La Batalla en el GATT por la libertad del comercio del banano." In *Informe UPEB*, no. 95. Panama City, Panama: UPEB.

U.S./Labor Education in the Americas Project (US/LEAP). 2001. "EU-US Banana Agreement a Reprieve for Most Workers, Puts Impending Dole Campaign on Hold." Chicago: US/LEAP.

U.S. Trade Representative (USTR). 1994. "Analysis of January 17, 1994 GATT Panel Ruling Against EU Common Organization of the Banana Market." Washington DC: USTR.

———. 1999a. "USTR Announces Final Product List in Bananas Dispute." Washington DC: USTR, April 9.

———. 1999b. "WTO Authorizes U.S. to Retaliate in Banana Dispute." Washington DC: USTR, April 20.

———. 2001a. "U.S. Government and European Commission reach Agreement to Resolve Long-standing Banana Dispute." Washington DC: USTR, April 11.

———. 2001b. "The U.S.–EU Banana Agreement: Frequently Asked Questions." Washington DC: USTR, April 11.

Van Sant, Herman. 1993. "The Social and Economic Impacts of EC Banana Importation Regulations." Presentation to the "South America Today" seminar, Guayaquil, Ecuador, June 30.

Vix, Cecily. 1996. "Unrest in the Banana Belt: The Labour Movement of the *Banderas Unidas.*" Unpublished report prepared for the National Labor Relations Board. Belize City, Belize.

Welch, Barbara M. 1996. *Survival by Association: Supply Management Landscapes of the Eastern Caribbean.* Mona, Jamaica: University of the West Indies Press.

Wiley, James E. 1994. "Quite Possibly the World's Most Perfect Neocolonial Fruit: The Banana Enters a New Stage." *International Third World Studies Journal and Review* 6 (1): 65–74.

———. 1998. "Economic Diversification in a Mini-state: Dominica in a Neoliberal Era." In *Globalization and Neoliberalism: The Caribbean Context*, edited by Thomas Klak. Lanham MD: Rowman and Littlefield Publishers.

World Trade Organization (WTO). 2000. *Dispute Settlement Reports 1997*, Vol. 2. Cambridge, UK: Cambridge University Press.

WTO Secretariat. 1999. *Guide to the Uruguay Round Agreements.* The Hague, Netherlands: Kluwer Law International.

Personal Communications:
Names, Locations, and Dates of Interviews

Alegría, Doris. 2000. Administrative Assistant, Dirección Nacional del Banano, Ministry of Trade and Industry. Panama City, Panama, December 5.

Amatraesijot, S. 2001. Manager, Farm 4, Jarikaba Plantation. Saramacca District, Suriname, August 22.

Appleton, Zelie. 1996. EU Directorate General VI. Brussels, Belgium, June 3.

Belon, Valerie. 1998. Economic, Commercial, Political, and Public Affairs Officer, U.S. Embassy. Belize City, Belize, June 8.

Benoit, Oliver. 1997. Director, Diversification Unit, Grenada Ministry of Agriculture. St. George's, Grenada, January 11.

Betancourt, Roberto. 2001. Ambassador of Ecuador to the WTO. Geneva, Switzerland, January 25.

Boatswain, Anthony. 1997. General Manager, Grenada Industrial Development Corporation. Frequente Industrial Park, Grenada, January 9.

Bouflet, Jean-Jacques. 2001. Minister-Counselor for Legal Affairs, EU Delegation to the WTO. Geneva, Switzerland, January 25.

Bradley, Elizabeth, and Jerrol Moriah. 2001. Europe Section Specialists, Ministry of Foreign Affairs. Paramaribo, Suriname, August 23.

Brinard, Philippe. 1996. Delegate General, European Community Banana Trade Association. Brussels, Belgium, June 2.

Brown, Dr. Kathy-Ann. 2001. Legal Advisor, Caribbean Regional Negotiating Machinery. Geneva, Switzerland, January 24.

Bundel-Griffith, Lilian. 2001. Administrative Technician, Office of the General Secretariat of the OAS. Paramaribo, Suriname, August 17.

Clarendon, Hannah. 1995. General Manager, Dominica Export Import Agency. (DEXIA). Roseau, Dominica, May 26.

Cyrus, Conrad. 1995. Chairman / Director of the Dominica Banana Growers Association. Roseau, Dominica, May 30.

Durand, Tim. 1996. Public Relations Officer, Dominica Banana Marketing Corporation. Roseau, Dominica, March 15.

Fagin, Vantil. 1996. Proprietor, Bello, Ltd. Roseau, Dominica, March 13.

Flores, Zaid. 1998. Secretary General, Belize Banana Growers Association. Big Creek, Stann Creek District, Belize, June 19.

Forman, Aldos. 1998. Foreman, Big Creek Port Facility. Big Creek, Stann Creek District, Belize, June 18.

Garelli, Stéphane. 2003. Director, *World Competitiveness Yearbook*. Lausanne, Switzerland, May 28.

Gomez, David. 2001. Minister-Counselor and Permanent Representative of Belize to the WTO. Geneva, Switzerland, January 24.

Gregoire, Sheridan, and Michael Fadelle. 1996. General Manager and Senior Investment Promotion Officer, National Development Corporation. Roseau, Dominica, March 11.

Harris, Carey. 1995. Director of the Diversification Implementation Unit, Ministry of Finance and Development. Roseau, Dominica, June 2.

James, Heather. 1995. Chief Technical Officer for the Ministry of Trade, Industry, and Tourism. Roseau, Dominica, June 6.

Jean-Pierre, Tony. 2000. Communications Assistant, St. Lucia Banana Corporation. Castries, St. Lucia, May 29.

Kieswetter-Alemán, Arq. Diana. 2000. Director, Dirección Nacional del Banano, Ministry of Trade and Industry. Panama City, Panama, November 29.

Livingston-Andall, Lucia, and Gregory Renwick. 1997. Director and Trade-Investment Advisor, Division of Trade, Industry, and Energy, Ministry of Finance. St. Georges, Grenada. January 9.

Magloire-Apgar, Sonia. 1995. Permanent Secretary, Ministry of External Affairs. Roseau, Dominica, May 29.

Mark, John. 1997. Acting General Manager, Grenada Banana Cooperative Society. St. George's, Grenada, January 17.

Martin, Urban. 1995. IICA Coordinator for Dominica. Roseau, Dominica, May 22.

Martinez, Edwin. 1998. IICA Coordinator for Belize. Belmopan, Belize, June 16.

Matthews, Chris. 1993. Press Officer, Delegation of the European Communities. New York, December 3.

McDonald, Frank. 1995. Research Associate, Caribbean Agricultural Research and Development Institute (CARDI). Roseau, Dominica, May 25.

Montero, Antonio. 1994. Director, Instituto Centroamericano de Asesoría Laboral. San Jose, Costa Rica, January 27.

Parham, Harold. 1998. Principal Agricultural Officer (Extension), Belize Ministry of Agriculture. Belmopan, Belize, June 16.

Reid, Errol. 2000. Director of Technical Services, WIBDECO. Castries, St. Lucia, May 25.

Reid, Richard. 1998. Senior Trade Economist, Belize Ministry of Trade and Industry. Belmopan, Belize, June 16.

Robinson, Dr. Donovan. 1995. Chief Technical Officer, Dominica Ministry of Agriculture. Roseau, Dominica, May 23.

Roches, Dean. 1998. Programme Support Director, SPEAR. Belize City, Belize, June 10.

Saborio Soto, Ronald. 2001. Ambassador of Costa Rica to the WTO. Geneva, Switzerland, January 24.

Sahtoe, Jaswant. 2001. Acting Permanent Secretary, Ministry of Agriculture, Animal Husbandry, and Fisheries. Paramaribo, Suriname, August 21.

Sánchez, Alex. 1993. Business Advisor to Costa Rican Consulate. New York, December 21.

Satney, Andrew. Marketing Intelligence Officer, Agricultural Diversification Co-ordinating Unit (ADCU) of the OECS. Roseau, Dominica, May 25, 1995.

Suescum, Alfredo A. 2001. Ambassador and Permanent Representative of Panama to the WTO. Geneva, Switzerland, January 23.

Van der Ploeg, J. L. J. 2001. Project Advisor, EU Commission Delegation in Suriname. Paramaribo, Suriname, August 21.

Vlijter, Ronny. 2001. Assistant Director, Surland Company. Jarikaba Plantation, Saramacca District, Suriname, August 22.

Woods, Valerie. 1998. Deputy Director of the Belize Tourism Board. Belize City, Belize, June 10.

Zúñiga, Martín. 1994. Economic Analyst, CORBANA. San José, Costa Rica, January.

Index

Fyffes Ltd.: 2001 Council Regulation and, 235; Banana Framework Agreement and, 186; Belize and, 100, 102–3, 104–5; Caribbean banana industry and, 14, 75–76, 80–81; licensing systems and, 135; Suriname and, 113–14

GATT. *See* General Agreement on Tariffs and Trade (GATT)

Geest Industries Ltd., 78, 80–81, 87; Banana Framework Agreement and, 186–88; basic peasant farmer production system and, 91–94; licensing systems and, 135; relationship of with banana farmers, 89–90

General Agreement on Tariffs and Trade (GATT), xi, 142; Banana Framework Agreement and, 183–85; complaints against the EU and, 182–83; European Union ordered to comply with, 182–83; Kennedy Round and, 166; and Marrakech Accords, 167, 170–71, 173–74; origins and experience of, 164–66; and Tokyo Round, 166; and Uruguay Round, 166–74; WTO waivers and, 175–76. *See also* World Trade Organization (WTO)

geography of the banana industry, xxi–xxii, *xxxi*, 138–39; Belize and, 99–100, *101*; Ecuador and, 40–42; Latin American, 5, 23–29;

shifting, 37–38; Suriname and, 108–10; Windward Islands and, 73, 79

Germany, 125, 131, 139, 144, 195

Ghana, 155

globalization: agricultural, xi–xiii; associations of, xi, xix; banana industry and, xii–xiii, xviii–xxii, 29–33, 84, 86, 121–22; diversification and, 215–24, 228–30; neoliberalism and, 240–45. *See also* banana war, U.S.-EU; Single European Market (SEM)

gracias law, 17

Great Depression, the, 35–40; and labor issues in Latin America, 43–44

Greece, 126, 130

Grenada, 78–81, *83, 85*, 96; Council Regulation 404/93 and, 142; diversification in, 219, 229–30; Lomé Convention and, 157. *See also* Caribbean banana industry growers associations, banana, 86–88, 90, 104, 251n5

Guadeloupe, 116–21

Guatemala, 13–14; civil war in, 65; complaints to the WTO by, 190–93; Council Regulation 404/93 and, 142, 145; Del Monte Corporation in, 50–51; GNP of, 38, 40; land acquisition in, 15–17; Panama Disease in, 37–38; politics in, 49–50; railroads in, 8–9, 19–20; ships in, 20–21; tariffs in, 250n2

151, 156–58, 242; complaints about, 191–92; funding by, 153–54; goals of, 152–53; mechanisms of, 152–54, 175; origins and philosophy of, 148–51; replaced with the Cotonou Agreement, 158–61; time periods and meetings of, 147–48, 150–51

marketing of bananas, xvii, 64; licensing systems and, 134–37; and niche market strategy, 224–28
Marrakech accords, 167, 170–71, 173–74
Martinique, 116–21. *See also* Caribbean banana industry
Mexico, 31–33, 224; complaints to the WTO by, 190–93
most favored nation (MPN) principle, 169, 174, 193
Multinational Banana Marketing Company. *See* COMUNBANA

nationalism, economic, 55–56
native origin of bananas, 3–4
neoliberalism, 240–45
Netherlands, the, 127, 139, 144. *See also* Suriname
Nicaragua, 57–60, 65, 142; Banana Framework Agreement and, 183–85
niche market strategy, 224–28
Noboa, 67–68; Banana Framework Agreement and, 186
nontraditional agricultural exports, 207–8

North American Free Trade Agreement (NAFTA), xi, xix

Organization of American States (OAS), 57
Organization of Petroleum Exporting Countries (OPEC), 56

packing, banana, 92–94, 107–8, 112, 251n6, 252n3
Panama, 13–14; GNP of, 38, 40; Panama Disease in, 37–38; railroads in, 11–12; response to Council Regulation 404/93, 195; Union of Banana Exporting Countries and, 60–63
Panama Accord of 1974, 60
Panama Disease, 37–38, 45–47, 48, 78
Partido Unido del Socialismo Cristiano (PUSC), 67
peasant farmers, 81–86; and production system, 91–94
perishable nature of bananas, 4–5, 92–93
Portugal, 126
Prebisch, Raul, 56

quotas, 183–85, 193–96, 246, 253n3; competitiveness and, 213–14; first-come, first-served controversy and, 199–203

railroads and the banana industry, 6–12, 19–20, 42; land acquisitions by, 15–17
Reagan, Ronald, administration of, 65

United Brands, 52, 57

United Fruit Company (UFCO), 5, 11; associate producers and, 47–49; Chiquita and, 31; commissaries and, 21–22; competitors of, 49–53; and development of the banana empire, 29–33, 35; founding of, 12–14, 30–31, 249n2; geographic isolation and, 23–29; Jamaica and, 73–75; labor issues and, 42–45; in Mexico, 31–33; Panama Disease and, 45–47; ships and, 20–21; vertical integration by, 18–23

United Kingdom, the, 144; Belize and, 100, 102–3, 104–5; colonialism and, 75–78; Lomé Convention and, 149; and origins of the eastern Caribbean banana industry, 78–81; Suriname and, 113–14

United Nations: Conference on Trade and Development, 221; Economic Commission for Latin America, 56; and Food and Agriculture Organization (FAO), 57, 61

United States: antimonopoly activity of, 50; and banana war with the EU, xii, xvii–xviii, 188–93; complaints to the WTO by, 190–93; introduction of the banana to, xvii; niche markets in, 225; reaction by to the 2001 Council Regulation, 234–36; sanctions levied by, 196–203. *See also* banana war, U.S.-EU

UPEB (Union of Banana Exporting Countries). *See* Union of Banana Exporting Countries (UPEB)

Urubá region, Colombia, 249n2

Uruguay Round, GATT. *See* General Agreement on Tariffs and Trade (GATT)

Vaccaro Brothers Company, 10, 249n2

Venezuela, 56, 142; Banana Framework Agreement and, 183–85

vertical integration, xx, 18–23

wages in the banana industry, 105–6, 114, 119, 202

waivers, WTO, 175–76

West Indies Fruit Company, 50

Windward Islands, 73, 79, 81, 94, 97, 105, 243–44; Banana Development Corporation (WIBDECO) and, 88, 186–88; Banana Growers Association (WINBAN) and, 88, 91; Crop Insurance Program and, 87. *See also* Caribbean banana industry

World Competitiveness Project, 210

World Competitiveness Yearbook, 209

World Trade Organization (WTO), the, xi, xviii, 160; agricultural trade and, 171–74; complaints against the EU made to, 190–93; dispute settlement by, 176–80, 236–40; future role of, 246; General Council and, 169–70;

In the At Table series

Spiced
Recipes from Le Pré Verre
Philippe Delacourcelle
Translated and with a preface by Adele King and Bruce King

A Sacred Feast
Reflections on Sacred Harp Singing and Dinner on the Ground
Kathryn Eastburn

Eating in Eden
Food and American Utopias
Edited by Etta M. Madden and Martha L. Finch

Recovering Our Ancestors' Gardens
Indigenous Recipes and Guide to Diet and Fitness
Devon Abbott Mihesuah

Dueling Chefs
A Vegetarian and a Meat Lover Debate the Plate
Maggie Pleskac and Sean Carmichael

A Taste of Heritage
Crow Indian Recipes and Herbal Medicines
Alma Hogan Snell
Edited by Lisa Castle

The Banana
Empires, Trade Wars, and Globalization
James Wiley

Available in Bison Books Editions

The Food and Cooking of Eastern Europe
Lesley Chamberlain
With a new introduction by the author

The Food and Cooking of Russia
Lesley Chamberlain
With a new introduction by the author

The World on a Plate
A Tour through the History of America's Ethnic Cuisine
Joel Denker

Masters of American Cookery
M. F. K. Fisher, James Beard, Craig Claiborne, Julia Child
Betty Fussell
With a preface by the author

Good Things
Jane Grigson

Jane Grigson's Fruit Book
Jane Grigson
With a new introduction by Sara Dickerman

Jane Grigson's Vegetable Book
Jane Grigson
With a new introduction by Amy Sherman

Dining with Marcel Proust
A Practical Guide to French Cuisine of the Belle Epoque
Shirley King
Foreword by James Beard

Pampille's Table
Recipes and Writings from the French Countryside from Marthe Daudet's
Les Bons Plats de France
Translated and adapted by Shirley King

Moveable Feasts
The History, Science, and Lore of Food
Gregory McNamee

CPSIA information can be obtained at www.ICGtesting.com
Printed in the USA
LVOW05s2023060813

346393LV00004B/257/P